Defoe Abbott Marx Hardy Machiavelli Montaigne Haggard Chesterton Cooper Emerson Joyce Austen

Melville Hugo Eliot Grimm

Stoker Carroll Christie Maupassant Molière Byron Schiller Smith Kafka Hall

Wilde Garnett Engels Hawthorne

Goethe Einstein Fitzgerald Dostoyevsky Doyle Willis

Cotton Kipling Nietzsche Balzac

Baum Henry Flaubert Turgenev Crane

Leslie Dumas Stockton Vatsyayana Verne

Burroughs Gogol Vinci

Curtis Tocqueville Whitman Busch

Homer Widger Tolstoy Twain Scott Plato Harte

Darwin Thoreau

Potter Freud Zola Lawrence Dickens

Kant Jowett Stevenson Hesse Burton

Andersen Cervantes

London Descartes Voltaire Cooke

Poe Aristotle Wells

Hale James Hastings Irving

Bunner Shakespeare

Richter Chekhov Chambers da Shaw Benedict Alcott

Doré Dante Swift Wodehouse Pushkin Newton

tredition®

tredition was established in 2006 by Sandra Latusseck and Soenke Schulz. Based in Hamburg, Germany, tredition offers publishing solutions to authors and publishing houses, combined with worldwide distribution of printed and digital book content. tredition is uniquely positioned to enable authors and publishing houses to create books on their own terms and without conventional manufacturing risks.

For more information please visit: www.tredition.com

TREDITION CLASSICS

This book is part of the TREDITION CLASSICS series. The creators of this series are united by passion for literature and driven by the intention of making all public domain books available in printed format again - worldwide. Most TREDITION CLASSICS titles have been out of print and off the bookstore shelves for decades. At tredition we believe that a great book never goes out of style and that its value is eternal. Several mostly non-profit literature projects provide content to tredition. To support their good work, tredition donates a portion of the proceeds from each sold copy. As a reader of a TREDITION CLASSICS book, you support our mission to save many of the amazing works of world literature from oblivion. See all available books at www.tredition.com.

Project Gutenberg

The content for this book has been graciously provided by Project Gutenberg. Project Gutenberg is a non-profit organization founded by Michael Hart in 1971 at the University of Illinois. The mission of Project Gutenberg is simple: To encourage the creation and distribution of eBooks. Project Gutenberg is the first and largest collection of public domain eBooks.

Desserts and Salads

Gesine Lemcke

Imprint

This book is part of TREDITION CLASSICS

Author: Gesine Lemcke
Cover design: Buchgut, Berlin – Germany

Publisher: tredition GmbH, Hamburg - Germany
ISBN: 978-3-8472-2493-8

www.tredition.com
www.tredition.de

Copyright:
The content of this book is sourced from the public domain.

The intention of the TREDITION CLASSICS series is to make world literature in the public domain available in printed format. Literary enthusiasts and organizations, such as Project Gutenberg, worldwide have scanned and digitally edited the original texts. tredition has subsequently formatted and redesigned the content into a modern reading layout. Therefore, we cannot guarantee the exact reproduction of the original format of a particular historic edition. Please also note that no modifications have been made to the spelling, therefore it may differ from the orthography used today.

GESINE LEMCKE.

DESSERTS AND
SALADS

BY

GESINE LEMCKE

AUTHOR OF
THE EUROPEAN AND AMERICAN CUISINE, AND CHAFING-
DISH RECIPES
PRINCIPAL AND OWNER OF THE BROOKLYN AND NEW
YORK COOKING COLLEGES

"Eating is a Necessity,
But Cooking is an Art."

NEW YORK AND LONDON
D. APPLETON AND COMPANY
1920

Printed in the United States of America

3

PREFACE.

I ASK every one who may become possessed of this book to read the recipes herein contained carefully and thoughtfully before attempting the making of any of them, and also to observe the following instructions:

Weigh and measure all ingredients exactly, and have everything ready to mix before you commence.

If you measure your ingredients by means of a cup be sure you use one which holds half a pint.

Use neither more nor less of anything than the recipe instructs you, and be sure to have your fire just right, as also instructed by the recipe.

If at first success does not come to you do not despair, but persist in following the advice of the old adage: "Try, try again."

You should always bear in mind that honest work is never lost and that reward must come in the end.

5

Desserts and Salads.

SAUCES.

1.**Wine Chaudeau.—** Into a lined saucepan put ½ bottle Rhine wine, 4 tablespoonfuls sugar, 1 teaspoonful cornstarch, the peel of ½ lemon and the yolks of 6 eggs; place the saucepan over a medium hot fire and beat the contents with an egg beater until just at boiling point; then instantly remove from the fire, beat a minute longer, pour into a sauce bowl and serve with boiled or baked pudding.

2.**White Wine Sauce.—** Over the fire place a saucepan containing 2 cups white wine, 4 tablespoonfuls sugar, 3 whole eggs, the yolks of 4 eggs and the peel and juice of 1 lemon; beat the contents of saucepan with an egg beater until nearly boiling; then instantly remove and serve.

3.**Wine Cream Sauce.** — ½ bottle white wine, ½ teaspoonful cornstarch, 3 eggs (yolks and whites beaten separately), 4 tablespoonfuls sugar and the peel and juice of ½ lemon; put all the ingredients except the whites of eggs in saucepan; beat with an egg beater until just about to boil; then remove from fire; have the whites beaten to a stiff froth; add them to the sauce, beat for a minute longer and then serve.

4.**Claret Sauce.** — Over the fire place a lined saucepan containing ½ bottle claret, 3 or 4 tablespoonfuls sugar, 1 lemon cut into slices and freed of the pits, apiece of cinnamon and 1 small tablespoonful cornstarch mixed with water or wine; stir constantly until it comes to a boil; then strain and serve. Or boil 1 tablespoonful cornstarch in 1½ cups water, with piece of cinnamon and a few slices of lemon, for a few minutes; then remove from the fire; add ½ pint claret and sugar to taste.

6

5.**Bishop Sauce.** — Boil 2 ounces of sago in 2 cups water, with 1 tablespoonful fine minced or ground bitter almonds, apiece of cinnamon and the peel of 1 lemon; when sago is done strain it through a sieve, add 1½ cups claret, ¼ pound sugar and 1 teaspoonful of bishop essence.

6.**Madeira Sauce, No. 1.** — Set a small saucepan on the stove with the yolks of 3 eggs, 1 cup Madeira and 2 tablespoonfuls sugar; stir until it comes to a boil; then remove from fire and add by degrees 4 tablespoonfuls sweet cream, stirring constantly, and serve.

7.**Madeira Sauce, No. 2.** — Mix 1 tablespoonful flour with 1½ spoonfuls butter; add 1½ cups boiling water; boil 3 minutes, stirring constantly; remove from the fire, add ½ cup Madeira and 3 tablespoonfuls sugar.

8.**Butter Sauce.** — In a small saucepan mix 1 tablespoonful flour with a little cold water; add by degrees 1 cup of boiling water, stirring constantly; set the saucepan over the fire, add 1 heaping tablespoonful butter in small pieces; continue stirring and boil for a few minutes.

9.**Sherry Wine Sauce, No. 1.** — Add to the Butter Sauce ½ cup sugar and ½ pint sherry wine.

10.**Sherry Wine Sauce, No. 2.**— 1 cup sherry wine, ½ cup water, the yolks of 3 eggs, 2 tablespoonfuls sugar and the grated rind of ½ lemon; put all the ingredients in a small saucepan over the fire and keep stirring until the sauce begins to thicken; then take it off; if allowed to boil it will be spoiled, as it will immediately curdle; beat the whites to a stiff froth, stir them into the sauce and serve.

11.**Sherry Wine Sauce, No. 3.**— Melt in a small saucepan 1 tablespoonful butter; add 1 teaspoonful flour; when well mixed add 1 cup sherry wine, 2 tablespoonfuls sugar and the yolks of 4 eggs; stir briskly until the sauce is on the point of boiling; then instantly remove and serve with plum or bread pudding.

7

12.**Wine or Brandy Sauce.**— Prepare 1 cup Butter Sauce, sweeten it with sugar, add 1 glass brandy, port or sherry wine, alittle lemon juice and nutmeg.

13.**Arrack Sauce (Allemande).**— Mix 2 tablespoonfuls flour with some white wine; add in small pieces 2 tablespoonfuls butter, peel and juice of ½ lemon and 2 cups white wine; place a saucepan containing the ingredients over the fire and stir until it comes to a boil; remove from the fire, add 1 cup arrack and 1 cup sugar.

14.**Arrack Sauce (English).**— Put in a small saucepan 1 tablespoonful flour mixed with a little cold water, the yolks of 3 eggs, 1 tablespoonful butter, apiece of cinnamon, alittle lemon peel, 2 tablespoonfuls sugar and 1½ cups water; set saucepan over the fire, stir constantly until it commences to boil; then instantly remove from the stove, add a little lemon juice and ½ cup arrack. This sauce can be made with any kind of wine or brandy.

15.**Brandy Sauce (with Milk, "English Style").**— Put in a small saucepan 1 cup milk, the yolks of 2 eggs, 1 tablespoonful sugar and a little grated lemon peel; stir over the fire till the sauce is at boiling point; instantly remove and add 3 tablespoonfuls brandy; serve with plum pudding.

16.**Brandy Sauce (American), No. 1.**— Stir 4 tablespoonfuls powdered sugar with 1½ spoonfuls butter to a cream; add by degrees the yolks of 2 eggs, ½ cup boiling water and ½ cup brandy; put all the ingredients in a tin cup and set it in a saucepan of hot water; stir until the sauce is boiling hot; flavor with nutmeg and vanilla. This sauce may be made of wine in the same manner.

17.**Brandy Sauce, No. 2.**— Beat 1 tablespoonful butter with 6 tablespoonfuls powdered sugar to a cream; add by degrees 1 wineglassful of brandy, 3 tablespoonfuls boiling water and a little nutmeg; put the sauce into a tin cup, set in saucepan of boiling water and stir until the sauce is hot; but do not allow it to boil.

18.**Punch Sauce.**— Place a small vessel on the stove with 1 cup of rum, 2 tablespoonfuls powdered sugar, the grated rind of ½ 8 an orange and 1 teaspoonful vanilla essence; let it remain over the fire until the liquor catches a light flame; put on the lid for 1 minute; then remove it from the fire, add the juice of 1 orange and serve hot. This sauce is usually poured over the pudding.

19.**Rum Sauce.**— Mix ½ tablespoonful flour with a piece of butter the size of an egg; add 1 cup boiling water; when well mixed together add ½ cup Rhine wine, the peel and juice of ½ lemon, 4 tablespoonfuls sugar, apiece of cinnamon and the yolks of 3 eggs; place in a saucepan over the fire and beat with an egg beater till the sauce comes to a boil; instantly remove and add ½ cup rum. In place of rum, brandy may be used. NOTE.—The eggs may be omitted and 1 tablespoonful flour used instead of ½.

20.**Sauce à la Diaz.**— Place a tin pan over the fire with 1 cup rum, ½ cup Marella wine, 3 tablespoonfuls sugar, the grated rind of 1 orange and 1 teaspoonful vanilla; leave the pan on the stove until the liquor takes fire; then cover quickly; boil 1 minute; draw it from the fire to the side of the stove; let it stand a few minutes; then strain into a bowl; cover tightly and when cold pour it over the pudding.

21.**Wine Chaudeau (with Rum).**— Place a saucepan on the stove with 1 teaspoonful cornstarch mixed with a little cold water; add 2 whole eggs, the yolks of 2 eggs, 4 tablespoonfuls sugar, alittle lemon juice, some grated orange peel, ½ bottle Rhine wine and 2 glasses of rum, stir with an egg beater until just about to boil; then instantly remove from the fire, stir for a few minutes longer and serve. Any other kind of liquor may be used instead of rum.

22.**Wine Sauce (with Almonds and Raisins).**— Put a small vessel over the fire with ½ bottle claret, 3 tablespoonfuls ground almonds,

3 tablespoonfuls raisins, apiece of cinnamon, 4 tablespoonfuls sugar and the peel of 1 lemon; stir until it boils; then remove from the fire, take out cinnamon and lemon peel and serve.

23.**Hard Sauce.**— Stir ¼ pound butter with 8 tablespoonfuls powdered sugar to a cream until it looks white; add by degrees 1 9 small glass of brandy (and, if liked, a little nutmeg); the yolks of 2 eggs may also be beaten through the sauce.

24.**Hard Sauce (with Cherries).**— Make a hard sauce with the yolks of 2 eggs and put some nice, ripe cherries (without the pits) into it; stir the whole well together and serve with suet pudding or dumplings. Blackberries, peaches or plums may be used instead of cherries.

25.**Strawberry Sauce.**— Boil in a saucepan 2 teaspoonfuls cornstarch in 1½ cups water with the rind of 1 lemon; take it from the fire, add 1 cup strawberry juice, alittle Rhine wine or claret and sweeten with sugar.

26.**Sauce of Apricots.**— Boil 3 tablespoonfuls apricot marmalade with 1 tablespoonful butter and ½ cup water 5 minutes; add 2 tablespoonfuls brandy and serve with boiled suet, batter pudding or apple dumplings.

27.**Sauce of Cherries, No. 1.**— Place in a saucepan 1 cup sugar, 1 cup water and ½ cup claret; when this boils add 1 pint of ripe cherries (without the pits); boil them 10 minutes; then take out the cherries and mix 1 teaspoonful cornstarch with a little water; add it to the sauce, boil a minute, strain and put cherries back into the sauce; serve cold.

28.**Sauce of Cherries, No. 2.**— Remove the pits from ½ pound ripe cherries; put the stones into a mortar and pound them fine; put them, with the cherries, 1 pint water and a piece of cinnamon, in a saucepan; add ¾ cup sugar and boil slowly ½ hour; strain and thicken the sauce with 2 teaspoonfuls cornstarch; boil a minute, add ½ cup claret and serve.

29.**Strawberry Hard Sauce.**— Stir 2 tablespoonfuls butter to a cream with 1 cup powdered sugar; add the yolks of 2 eggs; beat until very light and stir 1 cup nice, ripe strawberries through it; put the sauce in a glass dish, cover with the beaten whites of 2 eggs and put some nice strawberries on top of the sauce. Any other kind of fruit may be used instead of strawberries. Or stir ½ 10 cup butter with 1 cup powdered sugar to a cream; add the beaten white of 1 egg and 1 cup thoroughly mashed strawberries.

30.**Raspberry Sauce, No. 1.**— Put in a small saucepan the peel of 1 lemon, alittle piece of cinnamon, 1 cup water and 1 spoonful sugar; boil 5 minutes; mix 2 teaspoonfuls cornstarch with some cold water; add it to the contents of saucepan; boil a minute; add 1 cup raspberry juice or syrup and serve either hot or cold.

31.**Raspberry Sauce, No. 2.**— Set a saucepan on the stove with 1½ cups raspberry juice, ½ cup water, the juice and peel of 1 lemon, sugar to taste, 1 teaspoonful cornstarch and the yolks of 3 eggs; beat constantly with an egg beater until it comes to a boil; quickly remove it from the fire; beat for a few minutes longer; beat the whites of the 3 eggs to a stiff froth and stir them into the sauce.

32.**Huckleberry Sauce.**— Put the huckleberries with a little water in a saucepan over the fire; boil slowly for ½ hour; then strain through a sieve, sweeten with sugar and thicken with a little cornstarch; add a few tablespoonfuls port wine or a little lemon juice and claret; serve cold.

33.**Sauce of Dried Cherries.**— Wash 1 pound dried cherries; put them into a mortar and pound fine; place them in a saucepan with 3 or 4 cups water over the fire; add a few zwiebacks, apiece of cinnamon and boil 1 hour; strain through a sieve, add a little claret and lemon juice and sweeten with sugar.

34.**Nut Sauce.**— Stir 1 tablespoonful butter with 5 tablespoonfuls powdered sugar to a cream; add the yolks of 2 eggs and a few spoonfuls of water; put it in a tin pail; set in a vessel of hot water; stir until hot; remove the sauce from the fire, add ½ cup fine, minced almonds and flavor with vanilla. Fine, chopped, stoned raisins may be used instead of almonds.

35.**Hard Sauce (with Nuts).**— Prepare a hard sauce of 1 tablespoonful butter and 5 tablespoonfuls powdered sugar; beat this until white; add by degrees the yolks of 2 eggs; beat the whites of 2 eggs to a stiff froth; add the sauce gradually to the whites; beat 11 constantly with an egg beater; and lastly add 1 cup pounded or ground nuts, almonds, walnuts, hazel or hickory nuts. The nuts may be finely chopped if more convenient. This sauce may be prepared in the same manner with peaches, apricots (peeled and cut into pieces) or preserved pineapple.

36.**Strawberry Custard Sauce.**— Place a small saucepan on the stove with 1 pint milk, the yolks of 2 eggs and 2 tablespoonfuls sugar; stir constantly until it comes to a boil; instantly remove from the fire, flavor with vanilla and set it away to cool; then stir 1 cup strawberries into it; beat the whites of the 2 eggs to a stiff froth and put it on top of the sauce. This sauce is excellent with strawberry shortcake. NOTE.—Any kind of fruit may be substituted for strawberries.

37.**Fruit Sauce (not boiled).**— Stir 1 cup raspberry juice and 1 of currants with 8 tablespoonfuls sugar for 20 minutes; serve with cold puddings. Or boil 2 teaspoonfuls cornstarch in water for a few minutes; sweeten with sugar; thin it with raspberry, currant or cherry juice; add a little Rhine wine and serve with cold pudding. This sauce is exceedingly nice when made of strawberries with the addition of the juice of 1 orange and a little grated skin.

38.**Peach Sauce, No. 1.**— To be served cold. Pare and cut in halves ½ dozen peaches; stew them in sugar syrup; press them through a sieve; thicken them with a little arrowroot or cornstarch; boil a minute, add a little white wine and serve. Or boil the peaches (after they are peeled and free from the stones) in sugar syrup until tender; then take them out, put in a dish, cut each half into 4 pieces and pour the liquor over them; then serve with tapioca pudding.

39.**Peach Sauce, No. 2.**— Beat 1 tablespoonful butter with 4 tablespoonfuls powdered sugar to a cream; add the yolks of 2 eggs; beat until very light and creamy; then beat the whites of the 2 eggs to a stiff froth; add the sauce to them by degrees; keep on beating with an egg beater until all is well mixed together and stir 1 cup of fine, cut peaches through it; serve with boiled pudding.

12

40.**Sauce of Currants and Raspberries.**— Wash ½ pound red currants and raspberries; sprinkle with sugar and let them stand ½ hour; prepare a sauce the same as for Peach Sauce and stir the fruit through it.

41.**Cream Sauce (with Jelly), No. 1.**— Stir 1 cup currant jelly until smooth; add 1 cup rich, sweet cream and beat with an egg beater to a froth; add a little arrack rum or Cognac and serve with cold pudding.

42.**Cream Sauce (with Jelly), No. 2.**— Beat ½ cup fruit jelly and the whites of 2 eggs to a stiff froth and serve with cold pudding.

43.**Lemon Sauce, No. 1.**— Stir 1 tablespoonful butter with 4 tablespoonfuls powdered sugar to a cream; add by degrees 1 beaten egg, the juice and grated rind of ½ lemon, alittle nutmeg and 4 tablespoonfuls boiling water; beat the sauce thoroughly for 5 minutes; put in a tin pail and set in saucepan of hot water; stir constantly until very hot, but do not allow it to boil.

44.**Lemon Custard Sauce.**— Place a saucepan with 1 pint milk, 3 whole eggs and 3 tablespoonfuls sugar over the fire and stir until it just comes to the boiling point; quickly remove, pour sauce into a dish, flavor with lemon essence and serve cold with cold pudding.

45.**Lemon Sauce (with Liquor).**— Melt in a saucepan 1 tablespoonful butter; add ½ tablespoonful flour; when well mixed pour in 1 cup boiling water; boil 2 minutes; remove from the fire, pour sauce into a bowl; add the juice of ½ lemon, alittle nutmeg and a glass of brandy; sweeten with sugar and serve hot. Very nice with rolly-poly pudding or apple dumplings. Sherry or Madeira wine may be used instead of brandy.

46.**Sauce à l'Orange.**— Stir the yolks of 4 eggs with 2 tablespoonfuls powdered sugar to a cream; add by degrees 1 cup sweet cream and stir constantly; add the grated rind of 1 orange; put the whole in a tin cup or pail, set in a vessel of hot water and stir all 13 the time until it is on the point of boiling; then instantly remove from the fire, strain through a sieve over the pudding and serve hot.

47.**Sauce au Kirsch.**— Boil 1 teaspoonful cornstarch in 1 cup water; sweeten with 2 tablespoonfuls sugar; add 2 tablespoonfuls kirsch and serve.

48.**Lemon Sauce, No. 2.** — Mix 2 teaspoonfuls flour with a little cold water; put it in a saucepan; add 1 pint boiling water, 1 tablespoonful butter and ½ cup sugar; stir until the sauce boils; then remove from the fire, add the juice of 1 lemon and a little of the grated rind and nutmeg.

49.**Lemon Cream Sauce.** — Put in a tin pail or cup 1½ cups milk, the yolks of 2 eggs and 2 tablespoonfuls sugar; set in a vessel of hot water; beat with an egg beater until the sauce comes to a boil; remove from the fire; add ½ teaspoonful lemon essence; beat the whites to a stiff froth and stir them into the sauce.

50.**Almond Sauce.** — Remove the brown skin of 2 ounces of almonds, ground or chopped fine; put them in a saucepan with 2 cups milk, 2 tablespoonfuls sugar, the yolks of 3 eggs and 1 teaspoonful of arrowroot; put the saucepan in a vessel of hot water; keep stirring until the sauce comes to a boil. Instead of almonds almond essence may be used; alittle brandy may also be added if liked.

51.**Chocolate Sauce.** — Boil ¼ pound grated chocolate with 1 cup water and 3 tablespoonfuls sugar for 5 minutes; beat up the yolks of 3 eggs with 1½ cups cold milk; add it to the chocolate; keep stirring until the sauce comes to a boil; instantly take it from the fire, beat for a few minutes longer and pour it into a sauce bowl; serve cold with cold pudding.

52.**Chocolate Cream Sauce.** — Boil ¼ pound grated chocolate with 1 cup water for 5 minutes; add sugar to taste; beat up the yolks of 3 eggs with 1½ cups sweet cream; add it to the chocolate; keep stirring until nearly boiling; remove from fire, add some vanilla essence and the beaten whites of the 3 eggs.

14

53.**Vanilla Cream Sauce.** — Put in a saucepan 2 cups sweet cream, 2 or 3 tablespoonfuls sugar, 2 whole eggs and the yolks of 2 eggs; set the saucepan in a vessel of hot water; beat with an egg beater till the sauce just comes to the boiling point; then instantly remove from the fire; do not allow the sauce to boil; flavor with vanilla extract and serve cold.

54.**Vanilla Sauce.** — Put in a tin cup or pail 2 cups milk and 1 teaspoonful cornstarch; add the yolks of 3 eggs and 2 tablespoonfuls sugar; place the cup in a vessel of hot water; beat with an egg beater until it comes to a boil; instantly remove; pour the sauce into a sauceière; flavor with 1 teaspoonful vanilla and serve cold. Do not allow the sauce to boil or it will curdle.

55.**Sauce à la Cream (sweet).**— Put in a tin pail 2 cups milk, the yolks of 2 eggs, 2 tablespoonfuls sugar and 1 teaspoonful cornstarch; set in a vessel of hot water; stir constantly until it comes to a boil; instantly remove; flavor with vanilla; beat the whites of the eggs to a stiff froth; pour the sauce into a glass dish, spread the beaten whites over it and dust some powdered sugar over all.

56.**White Sauce.**— Boil 2 teaspoonfuls arrowroot in 1 pint milk; add 2 tablespoonfuls sugar and 1 teaspoonful lemon essence; beat the white of 1 egg to a froth and stir it through the sauce when cold.

57.**Cream Sauce (plain).**— Stir ½ tablespoonful butter with 4 tablespoonfuls powdered sugar to a cream; boil 1 tablespoonful flour in 1 cup of water; pour it slowly into the creamed butter; keep on beating until the whole is well mixed; flavor with 1 teaspoonful lemon essence and serve hot.

58.**Vanilla Sauce (plain).**— Put in a saucepan 1 pint milk, 1½ teaspoonfuls cornstarch, sugar to taste and stir over the fire until it boils; flavor with 1 teaspoonful vanilla essence and serve when cold.

59.**Vanilla Sauce (with Cognac).**— Stir 2 tablespoonfuls butter with 6 tablespoonfuls powdered sugar to a cream; add by 15 degrees 3 tablespoonfuls Cognac, sherry or Madeira wine and ½ cup boiling water; keep beating all the time; put this in a tin pail and set in a vessel of hot water; keep stirring until hot, but do not allow it to boil; remove from the fire and add 2 teaspoonfuls vanilla essence.

60.**Caramel Sauce.**— Put 2 tablespoonfuls sugar in a saucepan over the fire; let it get light brown; add a little water; boil for a minute or two; then pour it into a small saucepan; add 1½ cups of milk or cream and the yolks of 2 eggs; set the saucepan in a vessel of hot water; stir until it comes to a boil; remove from the fire and flavor with 1 teaspoonful vanilla.

61.**Coffee Cream Sauce.**— Pour 2 cups boiling hot cream over 2 tablespoonfuls freshly ground coffee; cover tightly and let it stand 10 minutes; then strain the cream through a fine sieve; put the cream in a small saucepan; add the yolks of 2 eggs, 1 teaspoonful cornstarch and 2 tablespoonfuls sugar; put this over a moderate fire and stir until it comes to a boil; remove from the stove, pour it into a sauce bowl and stir the beaten whites of the eggs through it; serve cold.

62.**Nutmeg Sauce.**— Mix 1 tablespoonful butter with 1 tablespoonful flour; add 2 cups boiling water and boil 5 minutes; sweeten with sugar and flavor with grated nutmeg.

63.**Orange Cream Sauce.**— Stir the yolks of 4 eggs with 1½ tablespoonfuls sugar to a cream; add 1 teaspoonful butter, alittle grated orange peel and ½ pint sweet cream or milk; put the ingredients in a small saucepan over the fire and stir till boiling hot; when cold mix it with a few spoonfuls whipped cream. Lemon Sauce is made in the same manner. This sauce may also be flavored with vanilla or lemon extract.

64.**Sabayon Sauce.**— Put the yolks of 4 eggs and 1 whole egg in a lined saucepan and beat them with an egg beater to a froth; add 4 tablespoonfuls sugar, asmall piece of lemon peel, the juice of 1 lemon and ½ bottle of Rhine wine; 5 minutes before 16 serving put the saucepan over the fire and beat constantly till boiling hot; but do not allow it to boil; serve at once. Sabayon of Madeira or Malaga wine without lemon juice is made the same way. If rum is added in place of wine it is then called Rum Sabayon Sauce.

65.**Strawberry Chaudeau Sauce.**— Put 1 cup strawberry juice or syrup in a saucepan; sweeten to taste; add ½ cup white wine and the yolks of 2 eggs; beat this over the fire with an egg beater till it foams and rises up; remove from the fire and mix it with the beaten whites of 2 eggs; serve with vanilla koch or souflée.

66.**Pineapple Chaudeau Sauce.**— Put 1 cup pineapple juice or syrup in a saucepan; sweeten to taste; add ½ cup white wine and the yolks of 2 eggs; beat this over the fire with an egg beater till it foams and rises up; remove from the fire and mix it with the beaten whites of 2 eggs; serve with vanilla koch or souflée.

67.**Raspberry Chaudeau Sauce** is made the same as Strawberry Chaudeau Sauce.

68.**Cocoanut Snow Sauce.** — Beat the whites of 3 eggs to a stiff froth and boil 1 cup sugar with ½ cup water till it forms a thread between 2 fingers; then gradually pour it into the beaten whites, stirring constantly; next add 1 cup freshly grated cocoanut.

69.**Cocoanut Sauce (another way).** — Stir 2 tablespoonfuls butter with 1 cup powdered sugar to a cream; add by degrees the yolks of 2 eggs; then beat the whites to a stiff froth; mix them with the sauce; add ¾ cup freshly grated cocoanut and serve with boiled pudding.

70.**Snow Sauce (with orange flavor).** — Beat the whites of 2 eggs to a stiff froth; boil a small cup of sugar with ½ cup water till it forms a thread between two fingers; remove it from the fire; add the juice of 1 orange and gradually pour it while hot into the beaten whites, stirring constantly; add last a little grated rind of 17 orange and serve. Snow Sauce with lemon flavor is made the same way.

71.**Pistachio Sauce.** — Stir the yolks of 4 eggs with 1 pint sweet cream and 2 tablespoonfuls sugar over the fire till nearly boiling; remove from fire; add 2 ounces finely pounded pistachio nuts; serve when ice cold with frozen pudding.

72.**Cold Pineapple Sauce.** — Pare and grate a small, ripe pineapple; press it through a sieve; add 1 cup sugar and a glass of Rhine wine; let it stand on ice for 1 hour and serve with frozen pudding.

SYRUPS.

73.**Plain or Sugar Syrup.** — Dissolve 4 pounds white sugar, 1 quart cold water and the beaten white of 1 egg; stir until sugar is dissolved; simmer for 3 minutes; skim well, strain through a fine flannel bag and bottle in well corked bottles.

74.**Pineapple Syrup.** — Pare and cut some large, ripe pineapples into small pieces; put them in a stone jar or large bowl; sprinkle a little sugar between and let the pineapples stand covered with a cloth in cellar for 36 hours, or until they have bubbles on top; then strain through a sieve or coarse bag, and if not clear enough strain again through a flannel bag; add to each pint of juice 1 pound of sugar; stir until the sugar is melted; then put it over the fire and simmer 3 minutes; skim and put the syrup in bottles; cork well and keep them in cool place. This syrup may be thinned with 2 parts plain syrup.

75.Strawberry Syrup. — Choose none but fine, ripe berries if you wish your syrup to be good; mash the strawberries in a stone jar or bowl; cover with a thin white cloth and let them stand 24 hours at a temperature of 70° to 80° F.; then inclose in a flannel bag and press them; add to each pint of juice 1 pound sugar; stir until the sugar is dissolved; then put it over the fire, let it boil up, skim well, remove from fire and bottle while hot.

18

75a.Raspberry Syrup is made the same as strawberry.

76.Raspberry and Currant Syrup. — Take equal quantities of raspberries and currants; free the latter from stems; put the fruit together into a stone jar or bowl, mash it up, cover with a cloth and let stand for 24 hours; then inclose the fruit in a coarse bag, press out the juice and to each pint add 1 pound sugar; let it boil up and bottle.

77.Raspberry Syrup (without fruit). — To make 8 gallons of syrup prepare a plain syrup of 18 pounds sugar with 5 gallons of water and put it in a clean mixing barrel; next dissolve 2 ounces tataric acid in 1 pint cold water and add it to the syrup; then pour 1 quart boiling water over 4 ounces powdered orrisroot; let it get cold; then filter; add it also to the syrup and stir up well. Color it with the following mixture: Take ½ pound mallow or malva flowers and soak them in ½ gallon water for 6 hours; then mash in a mortar 2 ounces cochineal and 2 ounces alum and pour over these 2 quarts boiling water, and when cold filter; next mix both colors together, add them to the syrup and stir for 15-20 minutes. This is an excellent recipe for imitation of raspberry syrup.

78.Raspberry Syrup (without boiling). — Mash some ripe berries in a stone jar or bowl and set the paste for 3 days (covered with a linen cloth) in a cool cellar; then press out the juice through a coarse bag; let it stand for 6 hours; drain off the clear juice and leave the sediment; add to 1 pint juice 1 pound sugar, stir for 1 hour and bottle; cork bottles loosely and set them for 4 days in the sun; then filter through a fine flannel bag; re-bottle the syrup in small bottles, cork well and cover corks with beeswax. Syrup made in this way is excellent for sauces. Strawberry and Currant Syrup without boiling is made in the same manner.

79.**Blackberry Syrup.**— Mash the blackberries in a stone jar, cover and let them stand for 48 hours; then strain them through a bag; add to each pint of juice 1 pound sugar; stir until dissolved; put it over the fire to boil 3 minutes; skim well; add to each quart of syrup ½ gill of French brandy and bottle. Or take nice, ripe 19 berries, mash and strain them; add to each pint of juice 1 pound sugar, ½ teaspoonful ground cloves and the same of cinnamon and mace; boil 5 minutes; add to 1 gallon of syrup ½ pint brandy and bottle.

80.**Peach Syrup.**— Pare and cut the peaches into small pieces; put them in a preserving kettle with a little water; crack some of the peach stones, add them to the peaches and let boil slowly for 15 minutes; then strain through a flannel bag; add to 1 pint juice 1 pound sugar and boil a few minutes; skim well and bottle.

81.**Apricot Syrup** the same way. Or pare and cut the peaches into pieces, crack a few of the stones, add them to the fruit and let it stand 24 hours; then strain; allow for 1 pint juice 1 pound sugar; let it come to a boil; skim well and bottle.

82.**Cherry Syrup.**— Pound a sufficient quantity of ripe cherries (with the pits) in a porcelain or stone mortar; let it stand for 3 days; inclose them in a bag, press out the juice, add to each pint 1 pound sugar; let it boil up once, skim and put the syrup in bottles; cork and set away for use.

83.**Wild Cherry Syrup** is made in the same manner as the above.

84.**Wild Cherry Bark Syrup.**— Pour 1 pint cold water over 4 ounces well bruised wild cherry bark; let it stand for 36 hours; press out and let the liquid stand till clear; add 1½ pounds white sugar; stir until dissolved and strain through fine flannel bag; set away in well corked bottles.

85.**Vanilla Syrup.**— Add ½ ounce fluid extract of vanilla to 1 gallon plain syrup. Another recipe: Rub ½ ounce citric acid with a little plain syrup; add 1 fluid ounce extract of vanilla and 1 gallon plain syrup.

86.**Vanilla Cream Syrup.**— Add to 3 pints plain syrup 1 ounce extract of vanilla, 1 quart rich, sweet cream or condensed milk.

20

87.**Cream Syrup.**— 1 cup sweet cream, 1 cup milk and 1 pound sugar are well mixed together, and if it is to be kept for several days add a little bicarbonate of sodium.

88.**Lemon Syrup.**— Grate the rind of 16 large, fresh lemons over 8 pounds granulated sugar; add 2 quarts cold water and the juice of the lemons; stir until the sugar is melted; then strain through a fine flannel bag and put the syrup in well corked pint bottles. Be careful to grate off only the yellow part of the rind of the lemons; the white part will give the syrup a bitter taste. There is no better lemon syrup made than this. 2 to 3 tablespoonfuls of this syrup in a glass of cold water makes fine lemonade and is also excellent for mineral waters and sauces.

89.**Lemon Syrup (with Oil of Lemon).**— Add to 1 gallon plain syrup 25 drops oil of lemon and 10 drams citric acid; mix the oil and acid together gradually; then add the syrup slowly, and when well mixed bottle syrup and keep in a cool place for use.

90.**Another Recipe**:— Add to 1 gallon plain syrup 6 drams tartaric acid dissolved in a little warm water, 1 ounce gumarabic dissolved in 1 ounce warm water and ½ dram of the best lemon oil, or a sufficient quantity of lemon extract to flavor the syrup.

91.**Lemon Syrup (plain).**— Make of 8 pounds sugar and 2 quarts water a plain syrup; when nearly cold add 1 quart pure lemon juice; filter through a Canton flannel filter and bottle.

92.**Orange Syrup.**— Grate the rind of 12 oranges over 7 pounds granulated sugar; squeeze out the juice, strain and pour it over the sugar; add ½ gallon cold water; stir until sugar is dissolved; then strain through a fine flannel bag and bottle. Care should be taken to grate only the yellow part of the rind of the oranges, as the least particle of white will make the syrup bitter.

93.**Orange Flower Syrup.**— Add to 1 pint orange flower water 1¾ pounds sugar; stir until the sugar is dissolved; then bottle.

21

EXTRACTS AND ESSENCES.

94.**Essence of Lemon.**— Grate the rind of 12 lemons; put this in a bottle with 1 pint alcohol and 1 teaspoonful lemon oil; cork bottle tightly; set in a warm place; shake every day and after 2 weeks it will be ready for use.

95.**Essence of Vanilla.**— Take 1 ounce vanilla beans; split each bean in two (lengthwise); then cut into small pieces; put these into a large bottle with 1 pint alcohol and 1 pint water; cork the bottle, not too tightly; set in a warm place for 3 weeks and shake it once every day; it will them be ready for use.

96.Bischof **Essence.**— Pare off the peel of 12 green oranges; put them with 1 bottle of good rum in a glass jar that is used for preserving fruit; let it stand 24 hours; then pour the essence into small bottles and set in a cool place for further use; 2 tablespoonfuls to 1 bottle of claret are sufficient.

97.**Essence of Oranges.**— Pare off the peel of 8 yellow and 4 green oranges; put them in a large bottle or glass jar with 1 quart arrack; set in a warm place for 2 weeks; then strain through filtering paper, put into small bottles and set them in a cool place for further use; 2 tablespoonfuls essence are sufficient for 1 bottle wine.

98.**Peach Essence.**— Dissolve 1 fluid dram oil of bitter almonds in 7 pints rectified spirits of 90 per cent.; allow the solution to stand for a few days and then filter it; put away in well corked bottles.

99.**Bitter Almond Essence.**— Dissolve 1 fluid dram oil of bitter almonds in 3 quarts rectified spirits of 90 per cent. and store the fluid for some time before using it.

100.**Coffee Essence.**— Pour 3 pints rectified spirits of 90 per cent. over 5¼ ounces finely roasted and ground coffee; let it stand for several days, draw off the fluid and filter.

101.**Cherry Essence.**— Press out the flesh of ripe cherries; let the mass stand quietly in a moderately warm room until the pure 22 juice has separated from the pulp; then place the mass in a bag, press the juice out, let it stand for a few hours longer and add an equal quantity of rectified spirits of 90 per cent.

102.**Strawberry Essence.**— Bruise 4½ pounds wild strawberries; pour 3 quarts spirits of 90 per cent, over the mass; let it stand for some time and filter. The product will be about 1 gallon of strawberry essence.

103.**Raspberry Essence.**— Crush 2 pounds ripe raspberries; press them out and add 2 quarts rectified spirits of 90 per cent.

104.**Rose Essence.**— Dissolve 2 fluid drams rose oil in 1⅓ quarts rectified spirits of 90 per cent. and filter the solution.

105.**Orange Blossom Extract.**— Pour 1¼ pints boiling milk over 10½ ounces fresh orange blossoms; place same over the fire and let it boil up; then add 3 quarts rectified spirits of 90 per cent.; mix it thoroughly, add 2½ pints champagne and filter.

106.**Orange Peel Extract.**— Crush in a stone mortar the rind of 12 oranges with some sugar; place the mass into a glass jar; add ⅓ gallon of rectified spirits of 90 per cent.; let it stand for 4 days; then decant the clear liquid and filter it; put away in well corked bottles.

107.**Italian Meringue.**— Whites of 5 eggs beaten to a stiff froth, 1 pound sugar, 1 teaspoonful vanilla extract, ¾ cup water; put sugar and water over the fire in a saucepan (one of agateware is best); stir until sugar is dissolved; next put saucepan over the fire and boil till the sugar begins to foam; then take some of the boiling sugar in a spoon and blow it; if it flows off the spoon in large bubbles it is ready to use; wipe the rim of saucepan clean with a damp cloth and remove the sugar from fire; let it cool for 2 minutes; then pour it slowly into the beaten whites, stirring constantly.

108.**Meringue.**— ½ pound powdered sugar and the whites of 5 eggs; carefully separate the whites from the yolks; put the whites in a deep kettle for 15 minutes on ice; then whip it with an egg 23 beater to a stiff froth; mix in slowly the sugar and use at once. This meringue is used for ornamenting puddings and cakes.

109.**Spinach Green (for coloring).**— Wash a few handfuls spinach, press it out and pound in a mortar to a pulp; then press out the juice in a cloth; put the spinach liquid in a small saucepan; put it for a few minutes over the fire; as soon as the liquid curdles pour it on a fine sieve; let the water run off and the green which remains press through a fine sieve and put it in a well covered glass till wanted. Spinach green is used for coloring creams or puddings.

110.**Sugar Color.**— Place a saucepan with 1 pound sugar and ½ pint water over the fire and boil till the sugar is dark brown and nearly black; then add 1 pint boiling water; stir until all the sugar is well dissolved; boil it for a few minutes; then remove from fire and put it into a well corked bottle. This color is used for coloring soups, sauces and sometimes jellies.

111.**Lemon Sugar.**— Grate the rind of 12 lemons; mix the grated lemon peel with 1 pound powdered sugar; put into well closed jars and set in a cool place; is used for cake sauces and puddings instead of freshly grated lemon peel.

112.**Vanilla Sugar.**— Split the vanilla bean, lengthwise, in two; put some granulated sugar on a plate and scrape the seed out of the vanilla bean; mix it with the sugar and put away in a well closed jar.

113.**Red Sugar.**— Sift out all the fine part of ½ pound granulated sugar; put the sugar on a piece of thick brown paper, drop a few drops of cochineal over the sugar and rub it with the hands till the sugar becomes a red color.

114.**Green Sugar** is prepared in the same manner as the foregoing, but care must be taken to use only green vegetable coloring.

FRENCH CREAMS.

115.**Creme Française à la Vanille.**— Put 1 quart sweet cream with the yolks of 8 eggs into a saucepan; add ¾ cup sugar 24 and stir the whole over the fire with an egg beater till nearly boiling; remove from fire, add 2 teaspoonfuls essence of vanilla and 1½ ounces clarified gelatine (see Gelatine); continue stirring until the cream has cooled off; then set a plain form with tube in center into cracked ice, pour in the cream, cover and let it remain for 2 hours. If the form is oiled with fine almond oil the cream will turn out without dipping the form into hot water; the oiling is best done with a fine brush; the form is then turned upside down, so that all superfluous oil has a chance to run out.

116.**Creme Française au Chocolat.**— Melt ¼ pound grated chocolate in the oven; then put it with 1 quart cream, ¾ cup sugar and the yolks of 8 eggs over the fire; stir until nearly boiling; remove it from fire, add 1 teaspoonful essence of vanilla and 1 ounce clarified gelatine and finish same as in foregoing recipe.

117.**Creme Française aux Amandes.**— 1 quart cream, ¼ pound blanched and finely pounded almonds, the yolks of 8 eggs, ½ cup sugar and ½ vanilla bean; boil the cream and pour it over the pounded almonds; cover and let it stand till cold; then strain the cream through a sieve; place a saucepan with the cream, yolks, vanilla and sugar over the fire; stir with an egg beater till nearly boiling; remove it from the fire and finish the same as Crème Française à la Vanille.

118.**Creme Française au Café.**— Pour 1 quart boiling cream over 4 tablespoonfuls fresh, ground coffee; cover and let it stand for 5 minutes; then strain through a fine sieve of cloth; place a saucepan over the fire with the coffee cream, yolks of 8 eggs and 5 tablespoonfuls sugar and stir till nearly boiling, finish the same as Crème Française à la Vanille.

119.**Creme Française au Thé.**— Pour 1 quart boiling cream over 2 tablespoonfuls of the best black tea; let it stand, well covered, for 5 minutes; then strain; put the tea cream with the yolks of 8 eggs and 5 tablespoonfuls sugar in saucepan over the fire and stir till nearly boiling; finish the same as Crème Française à la Vanille.

25

120.**Creme Française au Marasquin.**— Place a saucepan with 1 quart cream, the yolks of 8 eggs, ½ cup sugar and 1 teaspoonful vanilla over the fire and stir till nearly boiling; remove it from the fire and add 16 sheets of gelatine which has been soaked in cold water for 10 minutes and pressed out; add lastly ½ pint maraschino and finish the same as Crème Française à la Vanille.

121.**Creme Française au Rhum** is made the same as the foregoing, substituting rum for maraschino.

122.**Petits Pots Creme à la Vanille.**— Mix well together 1 quart cream, yolks of 8 eggs, 4 whole eggs and 5 tablespoonfuls sugar; fill the cream into buttered custard cups, set them in a pan of hot water on top of the stove, cover with a pan or paper and boil till contents are firm; remove from fire and set aside to cool; in serving turn the cream onto a dish and send whipped cream to table with it.

123.**Creme au Bain-Marie au Caramel.**— Boil ¾ cup sugar to a caramel (see Boiling Sugar), add a little boiling water, remove from fire and stir for a few minutes; then place a saucepan with 1 quart cream or milk, 4 whole eggs and the yolks of 8 eggs over the fire; add the caramel sugar and stir till nearly boiling; fill the cream into a buttered form, cover it tightly and place the form into a vessel of hot water; let it stand for 1½ hours on the hot stove; the water should be boiling hot, but must not boil; when done take it from the fire, set in cool place; when cold and ready to serve turn the cream onto a round dish and send to table without sauce.

124.**Creme au Bain-Marie au Chocolat.**— Melt 4 tablespoonfuls Baker's chocolate in the oven; mix it with 1 quart cream or milk, 5 tablespoonfuls sugar and boil 5 minutes, stirring constantly; remove it from the fire and when cold add 4 whole eggs, the yolks of 8 and 1 teaspoonful vanilla; beat these well together with an egg beater; butter a form with a tube in the center and sprinkle with fine zwieback crumbs; pour in the cream, cover the form and set in a vessel of hot water; let it boil slowly for an hour; remove it from the fire and set aside to cool; when ready to serve 26 turn the cream onto a dish, garnish with fancy cake and send whipped cream à la vanilla to table with it.

125.**Vienna Orange Cream.**— Soak 1 ounce gelatine in ½ cup cold water 15 minutes, add ½ cup boiling water and stir over the fire till dissolved; stir the yolks of 12 eggs with 12 tablespoonfuls sugar to a cream; add by degrees the juice of 8 oranges and 3 lemons, and lastly the gelatine; continue stirring until it begins to thicken; then add the whites, beaten to a stiff froth; rinse out a mould with cold water and sprinkle with sugar, pour in the cream and set it on ice for 2 hours.

126.**Vienna Lemon Cream.**— Stir the yolks of 10 eggs with 1 cup sugar to a cream; add the juice of 4 lemons and the grated rind of ½ a one; lay 12 sheets of gelatine in cold water for 10 minutes, press out and dissolve it in ½ cup boiling water; add it by degrees to the cream and continue stirring till it begins to thicken; then add the whites, beaten to a stiff froth; rinse out a form with cold water and sprinkle with sugar, turn in the cream and set on ice till firm.

127. **Milk Cream.** — Soak 1 ounce gelatine in 1 cup of milk; place a saucepan with 3 cups milk, ¾ cup sugar, 1½ teaspoonfuls vanilla and the yolks of 6 eggs in a vessel of boiling water and stir with an egg beater till nearly boiling; remove from the fire, add the gelatine and stir till it becomes cold and begins to thicken; then add the whites, beaten to a stiff froth, turn the cream into a form and set on ice till firm; serve with cold pineapple or strawberry sauce; the form should be rinsed with cold water and sprinkled with sugar before the cream is put in.

128. **Russian Cream.** — Stir the yolks of 9 eggs with 9 tablespoonfuls sugar for ½ hour; add the juice and grated rind of 1 lemon and ½ pint best rum; lay 5 sheets of gelatine for 5 minutes in cold water, press out and put it in ½ cup boiling water; stir until dissolved; then mix it by degrees, stirring constantly, with the above mixture; when it begins to thicken add the whites, beaten 27 to a stiff froth, rinse out a mould with cold water, sprinkle with sugar, fill in the cream and set in a cool place till firm.

129. **Sabayon of Oranges.** — Soak 12 sheets of gelatine in cold water for 10 minutes; in the meantime put the juice of 4 oranges, the thin peel of 1, ¾ cup sugar, ½ bottle wine (white is best), 2 whole eggs and the yolks of 6 in a saucepan over the fire; beat this with an egg beater till nearly boiling; remove it from the fire and take out the peel, press out the gelatine, add it to the cream and continue the beating till cold; fill it into a cream form and place for 2 hours on ice.

130. **Sabayon of Lemon.** — Soak 12 sheets of gelatine in cold water; put in a saucepan 2 whole eggs and the yolks of 10; add 1 cup sugar, the juice of 3 lemons, the thin peel of 1 and ½ bottle Rhine wine; beat this with an egg beater till nearly boiling; remove at once, press out the gelatine, add it to the cream and continue beating till cold and beginning to thicken; fill it into a cream form and set on ice till wanted.

131. **Whipped Cream.** — Put 1 quart of rich, sweet cream into a deep vessel or stone jar and let it stand on ice for an hour; then beat it with an egg beater until stiff; then add sufficient powdered sugar to sweeten and any kind of flavor that may be liked.

132.Whipped Cream (with Strawberries).— Put 1 quart ripe strawberries in a colander and rinse with cold water; when well drained put the berries into a glass dish, sprinkle over them 1 cup powdered sugar and set for ½ hour on ice; whip 1 pint sweet cream to a froth, sweeten with powdered sugar and set on ice until wanted; when ready for use pour the cream over berries and serve at once, or send each in a separate dish to the table.

133.Cream (with Pineapple).— Prepare the cream in the same manner as the foregoing; pare a ripe sugar-loaf pineapple and break it from the stalk into pieces with a silver fork; put the fruit into a glass dish and sprinkle 1 cup sugar over it; set the dish on 28 ice for 1 hour; when ready to serve pour the cream over pineapple and send to table at once.

134.Whipped Cream (with Chocolate).— Boil ¼ pound grated chocolate in ½ cup water with ½ cup sugar and a little vanilla; when cold mix it with 1 pint whipped cream and set on ice till wanted.

135.Whipped Cream (with Oranges).— Pare 6 large oranges, cut them into pieces, remove the pits, put the fruit into a glass dish and sprinkle over it ½ cup powdered sugar; have 1 pint cream, beaten to a stiff froth, mixed with 3 or 4 tablespoonfuls powdered sugar and set fruit and cream on ice till wanted; when ready to serve pour cream over the oranges and send to table at once, or serve each in a separate dish.

136.Creme Fouettée à la Cobby.— Mix with whipped cream some fruit marmalade and put it in layers in a glass dish with some preserved cherries and macaroons between each layer; arrange the cream high up in the dish and garnish with lady fingers or fancy cake.

137.Whipped Cream (with Peaches).— Pare and cut 6 large, ripe peaches into quarters; put the fruit into a glass dish, sprinkle over it ½ cup powdered sugar and set the dish on ice for 1 hour; also have 1 pint of whipped and sweetened cream standing on ice; in serving cover the peaches with cream; break some lady fingers apart, stand them around the dish and serve at once. Or serve cream and fruit in separate dishes. Instead of fresh fruit preserved fruit may be used.

138.**Whipped Cream (with Cherries).**— Remove the pits from 1 pound of large cherries; put the fruit in a glass dish with ½ cup sugar; set the dish for an hour or two on ice; also have 1 pint of whipped cream on ice; when ready to serve spread the cream over the cherries, or serve each in a separate dish, and send sponge or fancy cake to the table with it. Canned cherries, apricots or peaches may be substituted for fresh fruit.

29

BOILED CREAMS.

139.**Vanilla Cream.**— Place a saucepan with 1 pint cream or milk over the fire, add 2 tablespoonfuls flour, the yolks of 4 eggs, 1 table-spoonful butter, 3 tablespoonfuls sugar, 1 teaspoonful vanilla extract and a sprinkle of salt; stir this until it comes to a boil; when cold mix cream with the yolk of 1 egg and a little sweet cream.

140.**Chocolate Cream.**— Mix 2 tablespoonfuls flour with 1 pint of milk or cream; add ½ teaspoonful vanilla extract, ¼ pound grated chocolate, 3 or 4 tablespoonfuls sugar, ½ tablespoonful butter, asprinkle of salt and the yolks of 4 eggs; place this in a saucepan over the fire and stir till it boils; then remove and set aside to cool.

141.**Orange Cream.**— Mix 2 tablespoonfuls flour with 1 pint milk or cream, add ½ tablespoonful butter, 3½ tablespoonfuls sugar, the yolks of 3 or 4 eggs and the grated rind of ½ an orange and stir the cream over the fire till it boils; then set aside to cool.

142.**Almond Cream.**— Pound 6 ounces of blanched almonds with a little cream to a paste; mix them with 1 pint of sweet cream or milk, add 2 tablespoonfuls flour, 3½ tablespoonfuls sugar, the yolks of 4 eggs, alittle salt, ½ teaspoonful vanilla extract and ½ table-spoonful butter; stir this over the fire till it boils; then remove and set the cream aside to cool.

143.**Coffee Cream.**— Put 3 ounces fresh roasted Mocha coffee into 1 pint boiling cream; let it stand 15 minutes; then strain, add to the coffee cream 2 tablespoonfuls flour, ½ tablespoonful butter, 4 tablespoonfuls sugar, asprinkle of salt and the yolks of 6 eggs; stir this over the fire till it boils; remove it and set the cream aside to cool.

144.**Creme Frangipane à la Vanille.**— Mix ½ cup flour with 2 cups cream, add 5 well beaten eggs, asprinkle of salt, 4 tablespoonfuls sugar, 1 tablespoonful butter and 1 teaspoonful vanilla; 30 stir this over the fire till it boils; remove it, add 8 well pounded macaroons and set aside to cool. Instead of vanilla a little orange or lemon peel may be substituted. Blanched almonds, raisins, currants, finely cut citron or any kind of fruit such as pineapples, strawberries or peaches may also be used.

JELLIES.

145.**Jellies** should be as clear as crystal, not too sweet and just firm enough to hold together. Jellies that have to stand any length of time on the buffets must, of course, be firmer. Agood plan is to make a trial by putting a little in a tin cup and setting it on ice before the jelly is put into a form.

146.**To Clarify Gelatine.**— Put 2 ounces gelatine in a saucepan, add ½ pint cold water and let it stand 10 minutes; then add ½ pint boiling water, set the saucepan in a vessel of boiling water and stir until gelatine is dissolved; beat the whites of 2 eggs to a froth, add the juice of 1 large lemon and a little cold water; stir this into the gelatine, continue stirring until it boils, remove to side of stove and let it stand 5 minutes without boiling; then strain through a jelly bag and use as directed in following recipes.

147.**To Clarify Sugar.**— Put 1 pound sugar in 1 pint cold water and stir till sugar is dissolved; then strain through a napkin. Aquicker way is to boil the sugar and water with the juice of 1 lemon for a few minutes and strain the same way.

148.**To Clarify Fruit Juice.**— Lay a few sheets of filtering paper in water and let them soak for 15 minutes, changing the water twice; then press them out, pick into small pieces, wet a little again with water and put the paper into a small sieve; pour the fruit juice onto the paper and let it run through into a dish. If not clear the first time pour back again and let it run through once more.

31

149.**Jelly Bag.**— Take ¾ yard of white flannel and make a bias bag; this is done by taking the flannel on the bias, sewing the bottom and side together to a point; cut it even on top and hem; then sew a string on each end of hem. In using the bag lay a broom with one end on the back of a chair and the other end on a table; tie the bag onto the broom, in the center, so that it hangs between the table and chair; set a bowl underneath the bag; then pour in the jelly; pour that which runs through first back again into the bag; repeat this once or twice more until the jelly runs through clear. When all the jelly has run through fill it into a mould and set either on ice or in a cool place.

150.**Orange Jelly.**— Clarify 2 ounces of gelatine as directed (see Clarifying Gelatine), dissolve 1 pound sugar in 1 pint water, add the thin peel of 2 oranges and let it stand 1 hour; then remove orange peel and strain the sugar syrup through a napkin; remove the peel from 4 oranges, divide them into small quarters and remove the pits without breaking the fruit; next pour the juice of 8 oranges and 2 lemons through filtering paper (see Clarifying Fruit Juice); as soon as the gelatine, fruit juice and sugar are clarified mix the three together, place a jelly form into cracked ice, pour in a few spoonfuls jelly and when firm lay in one-third of the orange quarters, which should be wiped dry with a napkin; add sufficient jelly to cover the fruit and when hard lay over another third; cover again with jelly and continue until all is used up; cover the form, lay some ice on top and let it stand till firm; when ready to serve dip the form into hot water, wipe it dry, remove cover, turn the jelly into a dish and serve with vanilla sauce or sweet cream. NOTE.—If the inside of jelly mould is brushed with pure almond oil the form need not be dipped in hot water, as the jelly will slip out without any trouble. Fine olive oil may also be used, but care should be taken to use only the very best, as otherwise the flavor of the jelly will be spoiled.

151.**Plain Orange Jelly.**— Dissolve and clarify 1 ounce gelatine in ½ pint water as directed, dissolve 1 cup sugar in ½ pint water, add the thin peel of 1 orange and let it stand 1 hour; then 32 strain through a napkin; let the juice of 5 oranges and 1 lemon run through filtering paper or a fine napkin; mix the gelatine, fruit juice and sugar syrup together, pour it into a jelly mould and set in a cool place to get firm; when ready to serve dip the form into hot water, turn the jelly onto a dish and serve with the following sauce:—Beat

1 egg to a froth, add by degrees 1 cup milk, 2 tablespoonfuls sugar and 1 teaspoonful vanilla extract.

152.**Lemon Jelly.**— Clarify 2 ounces gelatine as directed, dissolve 1¼ pounds sugar in 1 pint water, add the grated rind of 2 lemons and let it stand ½ hour; then strain through a napkin; let 1 pint of lemon juice run through filtering paper (see Clarifying Fruit Juice); when the three ingredients have been clarified mix them together, fill the jelly into a jelly mould, set it on ice or in a cold place to get firm and serve same as Orange Jelly.

153.**Strawberry Jelly.**— Put 1 quart ripe strawberries in a colander, rinse them off with cold water and when drained mash them well in a bowl with a silver spoon; dissolve ¾ pound sugar in 1 pint cold water, add the juice of 1 lemon and put it over the fire to boil 5 minutes; strain through a napkin and when cold pour it over the strawberries; let them stand 3 hours; then strain the berries, first through a jelly bag and then through filtering paper; also let the juice of 2 oranges run through filtering paper; clarify 2 ounces gelatine as directed and when cold add it to the fruit juice; then make a trial by filling a few spoonfuls in a tin cup and set it on ice to form; if not firm enough add a little more dissolved gelatine; fill the jelly alternately with large strawberries in a jelly form and finish the same as Orange Jelly; serve with whipped cream.

154.**Pineapple Jelly.**— Pare and cut a large, ripe pineapple into quarters, remove the hard core from the center and cut the quarters of pineapple into fine slices; dissolve 1 pound sugar in 1 pint cold water and juice of 1 lemon, pour it over the pineapple pieces, cover and let it stand for 2 hours; chop the eyes and hard core of pineapple very fine, put them with 1 pint water in a saucepan over the fire and boil slowly ½ hour; when cold strain them and add the 33 liquid to the pineapple; in the meantime clarify 2 ounces gelatine as directed; then drain the pineapple in a sieve, wipe the slices dry with a napkin and lay them on a plate; let the pineapple syrup run through filtering paper or napkin and mix it with the clarified gelatine; also let the juice of 2 oranges and 1 lemon run through filtering paper and add it to the jelly; then make a trial to see if firm enough; place jelly form in cracked ice, pour a few spoonfuls of jelly into the form and when hard put in a layer of pineapple; cover them with jelly and when firm put in another layer of pineapple; continue until all is used up; then cover the form, put some ice on top of form and let it remain till jelly is firm; serve with or without cream or vanilla sauce. This jelly may also be made of preserved pineapple.

155.**Jelly of Peaches.**— Pare 8 large, ripe peaches, cut them into halves, remove the stones and cut each half into 3 or 4 pieces; put the fruit into a bowl and pour over it 1 pint of sugar syrup; let them stand well covered for 2 hours; scald the pits, remove the brown skin and put them with the peaches; then let the syrup run through filtering paper, mix it with 2 ounces clarified gelatine, fill the jelly with the peaches and pits in alternate layers in a form and finish the same as Orange Jelly.

156.**Raspberry Jelly.**— Press the juice from 1 quart ripe raspberries, add the juice of 1 lemon and filter it through filtering paper (see Clarifying Fruit Juice); dissolve ¾ pound sugar in 1 pint water, strain through a napkin and add it to the raspberry juice; add 2 ounces clarified gelatine; set a jelly form into cracked ice and fill the jelly alternately with large, ripe raspberries into the form and finish the same as Orange Jelly.

157.**Wine Jelly.**— Soak 2 ounces gelatine in ½ pint cold water for 10 minutes; then add ½ pint boiling water and stir the whole over the fire till gelatine is dissolved; add the rind and juice of 1 lemon, 2 whole cloves, asmall piece of cinnamon and the well beaten whites of 2 eggs; stir this with an egg beater till it boils; 34 then remove the saucepan with its contents to side of stove and let it remain for 5 minutes without boiling; then strain it through a flannel jelly bag; dissolve ¾ pound sugar in 1 pint cold water, strain it through a napkin and add it with 1 pint Madeira to the gelatine; rinse out a jelly mould with cold water, pour in the jelly and set it on ice or in a cool place till firm. Instead of Madeira wine any other kind may be used.

158.**Rhine Wine Jelly.**— Dissolve and clarify 2 ounces gelatine and dissolve 1 pound sugar in 1 pint water; add the rind of 2 lemons and the juice of 1; let it stand 1 hour; then strain through a napkin; let the juice of 2 lemons run through filtering paper, add it with 1 pint Rhine wine and the sugar syrup to the clarified gelatine, fill the jelly in a form and set it on ice or in a cool place.

159.**Champagne Jelly.** — Dissolve and clarify 2 ounces gelatine (see Gelatine), dissolve ¾ pound sugar in 1 pint cold water, strain it through a napkin, add to the gelatine with ½ bottle champagne and the filtered juice of 4 lemons, fill into a form and set it in a cool place or on ice.

160.**Apple Jelly.** — Grate 1 quart tart apples, put them in a bag and press out the juice, add the juice of 1 orange and let both run through a filtering paper; clarify 2 ounces gelatine, dissolve ¾ pound sugar in 1 pint cold water, strain through a napkin and add it with the apple juice to the clarified gelatine; rinse a mould with cold water, pour in the jelly and set it in a cool place or on ice till firm. Another way is: — Pare, core and quarter some tart apples and boil them in sugar syrup to which the juice of 1 lemon has been added; when the apples are done remove carefully, so as not to break them, lay on a sieve to drain and when cold lay into the mould alternately with the jelly and finish like Orange Jelly.

161.**Cider Jelly.** — Soak 2 ounces gelatine in ½ pint cold water for 15 minutes; then add ½ pint boiling water, put it over the fire and stir till gelatine is dissolved; add the juice of 1 lemon and the beaten whites of 2 eggs; stir with an egg beater until it 35 boils; then draw to side of stove and let it stand 5 minutes; then strain through a flannel jelly bag; dissolve ¾ pound sugar in 1 quart sweet cider, strain through a jelly bag and add it to the gelatine; pour it into a jelly mould and set in a cool place until firm.

162.**Rose Jelly.** — Put 1 quart of freshly gathered rose leaves in a glass jar, squeeze the juice of 1 lemon over them, pour over the whole 1 cup boiling water, close the jar tightly and set aside till next day; then press out the juice (by putting the rose leaves in a coarse bag), let the liquid run through filtering paper (see Clarifying Fruit Juice), add 1 pint cold clarified sugar syrup, ½ pint white wine and 2 ounces clarified gelatine; next pour the jelly into a mould and set aside to cool.

163.**Gelée Russe.** — Clarify 1 ounce gelatine and dissolve ½ pound sugar in ½ pint water; add the grated peel of 2 lemons; let it stand 15 minutes; then strain through a napkin; let ½ pint lemon juice run through filtering paper; mix the clarified gelatine, sugar syrup and lemon juice together; put it in a deep kettle, set into cracked ice and whip the contents until it foams and begins to thicken; then fill it into a form and cover and pack with cracked ice till firm, which will take about 2 hours. Orange and Wine Jelly may be made in the same manner.

164.**Macédoine de Fruit à la Russe.** — Prepare a Rhine Wine Jelly, set a form into cracked ice, pour in a few spoonfuls jelly and let it get hard; lay over it a layer of fruit, such as strawberries, slices of pineapple or peaches, and pour over sufficient jelly to cover the fruit; put the remaining jelly into a deep kettle, set into cracked ice and beat with an egg beater till it foams and begins to thicken; then mix with 3 or 4 different kinds of fruit, either fresh or preserved, fill into the jelly form, cover closely and let it stand 2 hours; when ready to serve dip the form in hot water, wipe it dry, remove the cover, turn the jelly onto a dish and garnish with sugared fruit.

165.**Calvesfoot Jelly.** — Choose 4 calves' feet with the skin on (if without the skin 6 must be taken), crack and wash them 36 well, put over the fire, cover with cold water and boil till they fall apart; strain the liquor through a fine sieve and let it stand in a cool place; next day skim off every particle of fat and remove the sediment; put the jelly over the fire and reduce it down to 2 quarts by boiling; beat up the whites of 4 eggs, add a little cold water, the juice of 2 lemons and the thin peel of one, 6 cloves and a piece of cinnamon; add this to the contents of saucepan, stirring constantly; boil for a few minutes; then move the saucepan to side of stove and let it stand for 5 minutes without boiling; then strain it through a double flannel bag; dissolve 2 cups sugar in 1 pint sherry or Madeira wine, strain through a napkin and add it to the strained jelly; rinse out the moulds with cold water, put in the jelly and set in a cool place. This jelly may be put into tightly corked bottles, and will keep for a long time. When wanted for use set the bottle in hot water until the jelly melts; then pour it into moulds and set in a cool place till firm.

166.**Macédoine de Fruits au vin du Rhine.** — Prepare a Rhine wine jelly a little stiffer than the ordinary jellies; take large, ripe raspberries, strawberries, currants, peaches (pared and cut into eighths) and pineapples cut into small slices; put them in a dish on ice; next set a form into cracked ice, pour in a few spoonfuls jelly and when hard lay in some of the fruit, either each kind by itself in small clusters or mixed one with the other; pour over this sufficient jelly to cover the fruit; let it get hard and again lay over some fruit; continue alternately with fruit and jelly till form is full; cover and let it remain in ice till firm; in serving dip the form into hot water, wipe it dry and turn the macédoine onto a round dish. In winter preserved fruit and apples and pears may be used. The apples and pears are to be cut into quarters and boiled for a few minutes in sugar syrup. The latter should be colored with a little cochineal.

167.**Gelée à la Moscovite.**— Any kind of fruit jelly may be used for this, using only half the quantity of gelatine as for jelly; put into a form, cover it, paste a strip of buttered paper around the edge of cover and pack the form in ice and rock salt for 2 hours; 37 only freeze about an inch all around, leaving it soft in the center; preserved fruit may be mixed with the jelly before it is put into the form; serve the moscovite in a glass dish and garnish with fruit or fancy cake.

168.**Orange Baskets (with Jelly).**— Choose 1 dozen large oranges and cut them into the shape of small baskets with handles; this is done by holding the orange in the left hand and cutting with a penknife a small quarter from each side of the orange toward the top, so as to leave the skin for the handle ½ inch wide; then cut the skin evenly all around; next separate the inside from the outside skin with the penknife and completely hollow the orange out, so that only little more than half of the skin with the handle is left; cut the edges into small scallops with a scizzors and lay the baskets in cold water; press out the juice from the oranges and with it make a jelly (see Orange Jelly); take the baskets from the water, wipe dry and with a napkin under them set on a tray; have the jelly on ice and when it begins to thicken fill up the baskets and place them on ice; if there should be any small holes in the baskets paste them up from the outside with butter, which must be removed before serving; serve on a napkin and garnish with green leaves. These baskets may also be filled with Gelée Russe.

169.**Orange Quarters Used for Garnishing Jellies and Other Dishes.**— Take 6 large oranges, cut out a round piece on the side of stem and hollow out so that nothing is left but the outside skin; care must be taken to leave none of the white coating on the inside of skin; after preparing this way put them in a saucepan over the fire with boiling water and boil 5 minutes; rinse with cold water, wipe them dry and fill each one either with clear jelly of different colors or blanc-mange; set them on ice until hard; cut them into quarters and use for garnishing different dishes. Small patty forms filled with jelly are also used for the same purpose.

170.**Almond Blanc-Mange.** — Soak 1 ounce gelatine in 1 cup cold milk for 15 minutes; then add 3 cups boiling milk, 6 tablespoonfuls sugar, ¼ pound blanched almonds (among the latter there should be a few bitter ones) and pound them in a mortar with a little water to a paste; set the saucepan with its contents into a vessel of boiling water and stir till it boils; remove from the fire and let it stand for 5 minutes; then strain through a muslin bag, add 1 teaspoonful extract of vanilla and set aside to cool; rinse out a quart mould with cold water and sprinkle with sugar; pour in the cold blanc-mange and set in a cool place till it becomes firm; when ready to serve loosen the blanc-mange around the edge on top and turn it over onto a dish; it may then be served either with or without fruit or vanilla sauce. Instead of almonds any other kind of flavoring may be used.

171.**Chocolate Blanc-Mange.** — Soak 1 ounce gelatine in 1 cup cold milk for 15 minutes; then add 2½ cups boiling milk; mix ¼ pound grated Baker's chocolate with ½ cup cold milk; add it to the gelatine with 6 tablespoonfuls sugar; place this in a saucepan over the fire and stir till it boils; remove from fire, add 1 teaspoonful vanilla extract and when cold pour it into the moulds, which have been rinsed out with cold water and sprinkled with sugar; set in a cool place till firm; this may be served with or without vanilla sauce.

172.**Blanc-Mange Marbre au Chocolat.** — Make half the quantity of both the Almond and Chocolate Blanc-Mange; rinse out a mould with cold water, sprinkle with sugar and place into cracked ice; pour in a few spoonfuls almond blanc-mange and let it get firm; then put in a few spoonfuls chocolate blanc-mange; when the latter is firm again put in some of the almond blanc-mange; continue in this way until all is used; let the form remain for 2 hours on ice and then serve with vanilla sauce.

173.**Cream Blanc-Mange.**— Soak in a small tin ½ ounce gelatine in ½ cup cold water for 15 minutes; set the tin in a saucepan of boiling water and stir until gelatine is dissolved; beat 1 pint rich, sweet cream to a stiff froth; add 4 tablespoonfuls powdered sugar and 1 teaspoonful vanilla or lemon flavoring; when this is well mixed add the gelatine by degrees, beating constantly; rinse 39 out a mould with cold water, sprinkle with sugar, fill in the blanc-mange and set on ice an hour or two before serving.

174.**Plain Blanc-Mange.**— Boil 1 quart milk with 6 tablespoonfuls sugar; add 1 ounce gelatine which has been soaked in a little cold water for 15 minutes; stir this over the fire until gelatine is dissolved; remove it from fire and when cold add 2 teaspoonfuls vanilla; rinse out a form with cold water, sprinkle with sugar, pour in the blanc-mange and set it on ice; serve with vanilla sauce.

175.**Cocoanut Blanc-Mange.**— Stir into the plain blanc-mange when it begins to thicken 2 cups freshly grated cocoanut.

176.**Neapolitan Blanc-Mange.**— Prepare an almond blanc-mange, strain and divide it into 4 equal parts; add to first part 1 tablespoonful grated chocolate and let it boil for a few minutes; mix second part with the yolks of 2 eggs and stir it over the fire till just about to boil; add to third part a few drops of cochineal, to color it pink; leave fourth part uncolored; rinse out a mould with cold water, sprinkle with sugar and place it into cracked ice; as soon as the blanc-mange becomes cold and begins to thicken put in first the white; after 5 minutes put in the pink; again waiting 5 minutes, put in the yellow and after a few minutes put in the chocolate; let it remain on ice till firm; when ready to serve work top free from the edge with a few light touches of your finger and turn the blanc-mange onto a dish.

177.**Nest with Eggs.**— Prepare 1 quart almond blanc-mange; take 12 fresh eggs, make a small hole in one end of each and let the contents flow out; rinse each shell well with cold water; then fill them with blanc-mange and set in a pan of sugar or flour, the open end up; place them in a cool place till hard; boil 1 pound sugar to a crack and spin it into quite long threads (see Spinning Sugar); with these threads form a nest a little smaller than the dish it is to be served in; dip each egg into warm water, wipe dry, break shells from about the blanc-mange and lay the artificial eggs in the nest. Another way is to make 1½ quarts orange or wine jelly; cut the rind of 3 oranges into long narrow strips and boil them for 20 40 minutes in water, changing the water 3 times; drain them on a sieve; put 1½ cups sugar with 1 pint water over the fire and when it boils add the orange peel; boil 15 minutes; remove and drain them on a sieve; put half of the jelly into a glass dish and when firm lay the artificial eggs upon it; arrange them the same way that natural eggs are generally found in a nest; lay orange peel, which represents the straw, over and around the eggs; when the remaining jelly is cold and thick pour it over the eggs and set in a cool place to form.

178.**Fromage Bavarois à la Vanille.**— Soak 1 ounce gelatine in 1 cup cold water 20 minutes; place a saucepan with 1 pint cream, the yolks of 6 eggs, ½ cup sugar and 1 teaspoonful vanilla over the fire and stir till nearly boiling; remove it from the fire, add gelatine and stir till dissolved; set saucepan with its contents in a vessel of cold water and stir till it becomes cold and begins to thicken; then mix it with 1 pint whipped cream; rinse a form with cold water, sprinkle the inside with sugar, fill in the bavarois and set for 2 hours on ice; serve on a round dish garnished with fancy cakes.

179.**Fromage Bavarois à la Vanille, No. 2.**— Boil 6 tablespoonfuls sugar in 1 cup water 5 minutes and flavor with 1½ teaspoonfuls vanilla; soak 1 ounce gelatine in 1 cup cold water 15 minutes, add it to the boiling sugar syrup and stir till melted; then set aside; when cold and beginning to thicken mix it with 1 pint whipped cream and finish the same as in foregoing recipe.

180.**Fromage Bavarois** *aux Pistache*.— Chop or pound 6 ounces pistachio nuts and ¼ pound almonds as finely as possible, mix with 1 pint cold sugar syrup and let them stand 2 hours; then strain through a fine sieve, add a little spinach green (see Color) and 1 ounce dissolved gelatine; stir until it begins to thicken; then mix with 1 pint whipped cream; put this into a form and place on ice for 2 hours. This cream should have a delicate green color; it is served on a round dish.

181.**Fromage Bavarois aux Amandes.**— Scald ½ pound sweet and 10 bitter almonds with boiling water, remove the brown 41 skin and pound or chop them fine; place a saucepan over the fire with 1 pint milk, 6 tablespoonfuls sugar, the yolks of 6 eggs, 1½ teaspoonfuls vanilla and the pounded almonds; stir until nearly boiling; soak 1¼ ounces gelatine in 1 cup cold milk, add it to the hot milk and stir till dissolved; then strain through a sieve; when cold and beginning to thicken stir in 1 pint whipped cream, turn into a form and set for 2 hours in cracked ice.

182.**Fromage Bavarois au Café.**— Pour 1 pint boiling milk over 4 tablespoonfuls freshly ground coffee, cover and let it stand 5 minutes; strain through a fine cloth; soak 1 ounce gelatine in a little cold water 15 minutes and add it to the coffee milk with 6 table-spoonfuls sugar and the yolks of 6 eggs; stir this over the fire till it nearly boils; remove from the fire and when cold and beginning to thicken stir in 1 pint whipped cream, turn into a form and pack in cracked ice 2 hours.

183.**Fromage Bavarois au Thé.**— Pour 1 pint boiling milk over 2 tablespoonfuls of the best black or green tea, cover and let it stand 5 minutes; then strain and finish the same as Fromage Bavarois au Café.

184.**Fromage Bavarois au Chocolat.**— Boil 4 tablespoonfuls grat-ed chocolate in ½ pint water, add ½ cup sugar and 1 teaspoonful vanilla; soak 1 ounce gelatine in 1 cup cold water 15 minutes, add it to the chocolate and boil a few minutes; remove from the fire and when cold mix it with 1 pint whipped cream, turn into a form and pack in cracked ice for 2 hours; then serve on a round dish with vanilla sauce.

185.**Lemon Fromage.**— Dissolve 1 cup sugar in ½ pint water, add the thin peel of 1 lemon, the juice of 3 and boil 5 minutes; add 1 ounce gelatine which has been soaked in ½ pint cold water and stir it until dissolved; then strain and when cold and beginning to thick-en add 1 pint whipped cream; fill this into a form and place it on ice for 2 hours.

186.**Orange Fromage.**— Soak 1 ounce gelatine in 1 cup cold water 15 minutes; dissolve 1 cup sugar in 1 cup water, add the 42 thin peel of 1 orange and boil 5 minutes; add gelatine and stir till melted; mix it with the juice of 6 oranges, strain and when cold and beginning to thicken add 1 pint whipped cream; turn into a form and pack in ice for 2 hours.

187.**Pineapple Fromage.** — Soak 1½ ounces gelatine in 1 cup cold water 15 minutes and stir it over the fire till dissolved; take 1 can preserved pineapple, drain off the liquor and add it to the gelatine; when cold and beginning to thicken cut the pineapple into small dice; stir the fruit with 1 pint whipped cream into the gelatine, turn into a form and pack it in cracked ice for 2 hours. Or peel a large, ripe pineapple, remove the eyes and hard core, cut into small square pieces, put them in a dish, sprinkle over with 1 cup sugar and let them stand for 2 hours; chop the eyes and core fine and put them in a dish; boil ½ cup sugar with 1 cup water, pour it boiling hot over the chopped pineapple and let it stand till cold; soak 1½ ounces gelatine in ½ pint cold water, put it over the fire and stir till dissolved; strain the chopped pineapple through a fine sieve, drain off liquid from the pieces and add them together to the gelatine; set in ice and stir till it begins to thicken; then stir in the pineapple pieces and 1 pint whipped cream; fill it into a plain form with tube in center and pack in cracked ice and a little rock salt for 2 hours.

188.**Peach Fromage.** — Pare and cut into quarters 1½ dozen ripe peaches, put with 1 cup powdered sugar into a dish and let them stand 2 hours; also add the peach pits (after they have been scalded and freed from their brown skin); soak 1½ ounces gelatine in ¾ cup cold water for 15 minutes, add ¾ cup boiling water and stir over the fire till melted; strain and set aside to cool; press the peaches through a sieve, add gelatine and pits and stir till it begins to thicken; then carefully stir in 1 pint whipped cream, turn into a form and place for 2 hours on ice.

189.**Strawberry Fromage.** — Soak 1 ounce gelatine in ½ pint cold water 15 minutes; then stir it over the fire till dissolved; wash and press 1 quart fresh strawberries through a sieve, add 1 cup 43 powdered sugar, the gelatine and a few drops cochineal; stir until it begins to thicken; then add 1 pint whipped cream, turn into a form and pack for 2 hours in cracked ice and rock salt.

190.**Rum Bavarois.** — Soak 1¼ ounces gelatine in ¾ cup cold water 15 minutes, add ¾ cup boiling water, stir over the fire till dissolved, strain and set aside; place a saucepan with the yolks of 6 eggs, ¾ cup sugar and 1 pint milk over the fire and stir till nearly boiling; remove from the fire, add ½ pint rum and the gelatine and continue stirring until it begins to thicken; then stir in carefully 1 pint whipped cream, turn into a form and pack in cracked ice for 2 hours.

191.**Fromage Bavarois Cardinal.**— Soak ¾ ounce gelatine in ½ cup water 15 minutes; boil ¼ pound unsweetened grated chocolate in 1 cup water with 2 tablespoonfuls sugar; add the gelatine and stir till dissolved; lay a plain form into cracked ice, pour the chocolate in by degrees and keep turning so that chocolate may form a complete lining inside of form; then set the form straight and pour in the bottom the remaining chocolate; as soon as this is hard fill the form with Bavarois of Vanilla, No. 2, and let it remain buried in ice for 2 hours.

192.**Bavarois** may be made of different colors—such as pistachio cream outside and bavarois of almonds inside; or strawberries outside and vanilla bavarois inside.

193.**Fromage Bavarois au Pain Noir.**— Cut a small pumpernickel into slices, lay on a tin in the oven to dry and roll them fine; take 1 cup of these crumbs and stir them into a bavarois of almond or vanilla; after the cream has been added turn into a form and pack in ice for 2 hours. For all bavarois the forms may be lined first with jelly and decorated with fruits, nuts, currants, etc. In order to do this place a form into cracked ice and pour in a few spoonfuls fruit jelly; when firm take whatever is going to be used onto a larding needle, dip each piece into jelly and lay them into the form in fancy patterns; pour in a little more jelly and when firm lay the form over on its side; pour in a little jelly at a time; keep 44 turning form, so that the whole inside may be covered with jelly; then decorate the same as bottom and fill with Fromage Bavarois à la Vanille or any other kind.

194.**Snow Pudding.**— Soak 1 ounce gelatine in 1 pint cold water 20 minutes, add 1 pint boiling water, 1 cup sugar, the juice of 2 lemons and the thin peel of 1; set it over the fire, stir and boil a few minutes, strain through a sieve and when it begins to thicken add the beaten whites of 6 eggs; rinse out a form with cold water, sprinkle with granulated sugar, fill in the mixture and set in a cool place; when ready to serve turn the pudding onto a dish and serve with vanilla sauce made of the yolks of 6 eggs (see Sauce). Milk or cream may be substituted for water; then the lemon juice is omitted and lemon extract used for flavoring.

195.**Wine Pudding.**— Soak 1 ounce gelatine for 10 minutes in 1 pint cold water, add ¼ pound sugar, ½ pint red wine and ½ pint raspberry juice; stir over the fire till boiling hot, strain through a jelly bag and put in a form to cool; when firm turn out on a flat dish and serve with vanilla sauce or whipped cream.

196.**Apple Jelly Pudding.**— Boil 1½ pounds peeled apples with 1 quart water, stir through a sieve, add ½ pound sugar and the juice of 2 lemons; soak 15 sheets of white and 3 of red gelatine for 5 minutes in cold water, press out and mix with the apple sauce; stir over the fire until the gelatine is all dissolved; then pour into a form and set on ice to get firm; serve with vanilla sauce.

197.**Maraschino Pudding.**— Take 10 eggs, 10 tablespoonfuls sugar, 14 sheets gelatine (soaked in cold water), ¼ pint rum (or maraschino) and the peel and juice of 1 lemon; stir the yolks and sugar to a cream and add by degrees rum and lemon; press out the gelatine and dissolve in 1 cup boiling water; add it, stirring constantly, to the other mixture; add lastly the whites of the eggs, which have been beaten to a stiff froth; next pour into a mould and set aside to cool; the mould should be rinsed with cold water and sprinkled with granulated sugar before pouring the pudding into it.

45

198.**Manilla Pudding.**— Place a saucepan over the fire with 1 pint milk, the yolks of 5 eggs, 4 tablespoonfuls sugar and the peel of 1 lemon; stir this over the fire until just about to boil; then instantly remove; have 1 ounce gelatine soaked in 1 cup milk, which stir into the hot mixture and set aside to cool; as soon as it begins to thicken add the whites of the eggs, beaten to a stiff froth, pour into a mould and set on ice to get firm; serve with fruit or claret sauce; the mould should be rinsed with cold water and sprinkled with coarse sugar previous to being used.

199.**Rum Pudding.**— Take 10 eggs (yolks and whites beaten separately), 1 pint sweet cream, ½ pound sugar, ½ pint rum and 1½ ounces gelatine; stir the yolks of the eggs and sugar to a cream, add the cream and rum, put this in a tin pail and set in a vessel of hot water; keep stirring with an egg beater until just about to boil; then quickly remove from the fire; have gelatine soaked in a little cold water, add it to the cream and mix well; when cold add the beaten whites of the eggs, pour into a mould and set on ice; in serving turn out and send fruit sauce to table with it.

200.**Fine Chocolate Pudding.** — ¼ pound Baker's grated chocolate, 3 cups milk, 1 cup water, 1½ ounces gelatine, 5 tablespoonfuls sugar and 6 eggs; boil chocolate with the water until well dissolved; soak gelatine in a little cold water about 5 minutes; place a saucepan with the milk, sugar, 6 yolks of the eggs and the boiled chocolate over the fire; beat the whole with an egg beater until just about to boil; add the gelatine, remove from fire, continue beating for a little while longer and set aside to cool; when it begins to thicken add whites of the eggs, previously beaten to a stiff froth, and pour it into a jelly mould which has been well rinsed with cold water and sprinkled with sugar; set either on ice or in cold water to get firm. In serving turn pudding onto a glass dish and serve with the following sauce: — Place a saucepan over the fire with 2 eggs, 1 pint milk, 1 teaspoonful cornstarch and 2 tablespoonfuls sugar; stir with an egg beater until nearly boiling; quickly remove from fire, flavor with 1 teaspoonful vanilla extract and serve cold. This will make a sufficient quantity for a family of 10 persons.

46

201.**Fine Claret Pudding.** — 1 pint claret, ½ pint water, ½ tablespoonful cornstarch, the thin peel of ½ lemon, 4 tablespoonfuls sugar, 4 eggs and 8 sheets of red gelatine; lay gelatine in cold water and let it remain until the pudding is prepared; put the wine, water, cornstarch, sugar, lemon peel and yolks of the 4 eggs in a saucepan and beat it up well with an egg beater for 5 minutes; then place saucepan with its contents over the fire and continue beating till just before boiling; remove from the fire, squeeze the water from gelatine, put it into the saucepan and mix with its contents; then set aside to cool; as soon as it begins to thicken add the whites of the 4 eggs, previously beaten to a very stiff froth; when this is well blended together rinse a jelly mould with cold water, sprinkle with sugar, pour in the mixture and set it either in cold water or on ice to get firm; serve with vanilla or cream sauce or turn the pudding onto a glass dish and lay a border of whipped cream around it. This pudding if made according to above recipe is very fine and sufficient for a family of 6 persons.

202.**White Wine Pudding.**— ½ bottle white wine, 2 of red and 6 sheets of white gelatine, the grated rind and juice of 1 lemon, alittle vanilla, 5 eggs and 6 tablespoonfuls sugar; lay the gelatine in cold water; place a saucepan with yolks of the 5 eggs, lemon, sugar, vanilla and wine over the fire and stir constantly until just about to boil; then remove from fire, press gelatine out, add to the hot mixture and set aside to cool; as soon as it begins to set whip whites of the 5 eggs to a stiff froth and stir them through it; fill a jelly mould with the mixture and set it on ice to get firm; serve with vanilla sauce. The mould should be rinsed with cold water and dusted with coarse sugar previous to pouring the pudding into it.

203.**Cold Apple Pudding.**— Put 1½ pounds peeled and sliced apples in a saucepan with 1½ quarts water; stew till tender, strain through a colander, return it to saucepan and add 1 pound sugar; soak 2 ounces gelatine in a little cold water, add to the apples, let the whole boil for a few minutes and pour it into a form to cool; serve with vanilla sauce.

47

FINE COLD PUDDINGS.

204.**Pudding à la Polonaise.**— Beat the yolks of 10 eggs and 2 whole eggs with an egg beater with 1½ pints Rhine wine (or white wine), 1 cup sugar and the grated rind of 1 lemon and the juice of 4; strain this into a large kettle and beat over a slow fire till nearly boiling; remove the kettle, place it into cracked ice or cold water and continue beating till cold; in the meantime soak 1½ ounces gelatine in ½ cup cold water for 15 minutes, add ½ cup boiling water and stir over the fire till dissolved; then stir it slowly into the cream, beating constantly; add lastly ½ cup rum; next place a cream form into cracked ice, put in a few spoonfuls cream and put over this a layer of vanilla wafers which have been soaked in sugar syrup with a little rum; after 5 minutes add more cream and wafers; continue until the cream is used up; leave on ice for 2 hours; when ready to serve dip the form into hot water, turn the pudding onto a round dish and serve; sufficient for 12 persons. If this pudding is too large half the quantities may be used.

205.**Peach Pudding (with Champagne).—** Pare and cut into halves 1½ dozen large, ripe peaches; put them into a dish with the blanched pits, add 1 cup sugar, 1 teaspoonful vanilla, or put ½ stick vanilla between the fruit; cover and let them stand about 2 hours; then divide the peaches into 2 parts: press one part through a hair sieve and add the peach juice and 1½ ounces gelatine previously soaked in cold water and dissolved in boiling water; when this is well mixed set it aside; cut some small sponge cakes into slices, put on a plate and pour a little champagne over them; set a plain tin form into cracked ice and pour in some champagne jelly (see Jelly); let it get firm and lay in the center one of the peach pits; lay around this some of the peach halves, pour a few spoonfuls more jelly over them and then a thin layer of whipped champagne jelly which has been colored with cochineal to a delicate pink; add to the peaches which have been pressed through a sieve 1 pint whipped cream and ½ bottle champagne; fill the cream in alternate layers with peaches and sponge cake into the form; let the last layer 48 be cream; let the form remain 2 hours longer in the ice; in serving dip the form in hot water, turn the pudding onto a handsome dish and garnish the edge with small croutons of champagne jelly which has been colored to a delicate pink with cochineal. White wine may be substituted for champagne.

206.**Pineapple Pudding à la Royale.—** Pare and cut in half a nice, ripe pineapple; remove the hard part from the center and cut the pineapple into fine slices; put into a bowl and sprinkle 8 tablespoonfuls sugar over them, cover and let stand 2 hours; in the meantime prepare 1 pint white wine jelly; set a plain tin form into cracked ice, pour some jelly into it and let stand till firm; then put a wreath or a star of pineapple over the jelly, sprinkle a few blanched almonds between them and pour some more jelly over it; when this is firm turn form on its side, pour a little jelly in and keep turning in the cracked ice till jelly is firm; lay slices of pineapple on the sides, sprinkle blanched almonds cut into strips between, pour over a little more jelly and turn the form till all is firm; in the meantime boil 1 pound sugar with 1 cup water 10 minutes and add 1 ounce gelatine which has been previously soaked in ½ cup cold water and dissolved in ½ cup boiling water; remove the slices of pineapple, add the juice from pineapple to the boiled syrup, set this into cracked ice and stir till it begins to thicken; then add 1 pint whipped cream and fill the cream into the form alternately with layers of lady fingers and macaroons which have been previously dipped into the syrup; cover the form and pack it in ice for 2 hours; cut the remaining slices of pineapple into dice, mix with some of the cold jelly, put in small tin forms and garnish the pudding, when turned out, with them.

207.**Orange Pudding à la Maltaise.**— Boil 1½ cups sugar with 1 cup water 5 minutes; add the juice of 6 oranges, the grated rind of 2 and 1½ ounces gelatine which has been soaked for ½ hour in cold water; stir until gelatine is melted, strain through a fine sieve, place on ice and stir till it begins to thicken; then add 1 pint whipped cream; mix the juice of 6 oranges and 1 lemon with 1 cup sugar syrup and strain through a sieve; cut the crust off a sponge 49 cake which has been baked in a deep pan the day before, cut the cake into slices about ½ inch in thickness and dip each slice in the orange liquor; set a plain tin form into cracked ice and pour in ½ pint plain orange jelly (see Jelly); let this get firm; decorate the bottom with a wreath of green pistachio nuts or blanched almonds and currants, or any kind of fruit, such as strawberries, cherries or plums; pour over some jelly; as soon as firm add a few spoonfuls jelly, then a layer of the orange cream and over this the sponge cake; continue with layers of cream and sponge cake till all is used; let the last layer be cream; let the form remain in ice for 2 hours; in serving turn the pudding onto a handsome round dish and garnish with orange quarters glazed with sugar.

208.**Pudding de Savoie à l'Orange.**— Remove the skin from 3 oranges, divide them into quarters and remove pits without disfiguring the fruit; boil 1½ cups sugar with 1 cup water 5 minutes, remove it from the fire, add ½ pint Rhine wine, the juice of 6 large oranges and the grated rind of one: when cold add 2 ounces dissolved gelatine (see Gelatine), set on ice and stir till it begins to thicken; then add the orange quarters; place a tin form in cracked ice and cover the bottom with some clear orange or lemon jelly to the depth of about ½ inch; as soon as jelly is firm decorate the bottom with orange quarters and blanched nuts; add to the juice of 6 oranges ½ bottle Rhine wine and sweeten with sugar; cut a medium sized sponge cake into slices, dip in the orange juice and put them in alternate layers with orange and jelly into the form; let it remain on ice 2 hours; when ready to serve dip the form into hot water, turn onto a round dish and decorate the edge with orange quarters and finely chopped orange jelly.

209.Chestnut Pudding à la Dauphine.— Boil 1 pound chestnuts for a few minutes, throw them into cold water and remove outside and inside brown skin; then boil the chestnuts in milk till soft and press them through a sieve; add to purée the yolks of 6 eggs, 1 pint cream, 1 teaspoonful vanilla and 6 tablespoonfuls sugar; stir this over the fire till nearly boiling, add 1½ ounces dissolved gelatine, set the cream into cracked ice and stir till it begins 50 to thicken; cut some sponge cake into slices and pour a little rum over them; then place a tin form in cracked ice (if a form is not handy use a tin kettle), pour a few spoonfuls of the cream into it and let stand till firm; lay over this some preserved apricots or pineapples with ¼ pound citron cut into dice and the sponge cake; continue this in alternate layers till all is used; let the pudding remain for 2 hours in ice; when ready to serve dip the form into hot water, turn pudding onto a dish and pour ½ pint vanilla syrup over it.

210.Pudding à la Girot.— Place a saucepan on the stove with 1½ pints sweet cream, the yolks of 6 eggs, 4 tablespoonfuls sugar and 1 teaspoonful vanilla essence; stir this over the fire till nearly boiling, remove the cream and set aside to cool; then add 1½ ounces dissolved gelatine; soak ¼ pound lady fingers and ¼ pound macaroons in cherry wine; then place a tin pudding form with tube in the center into cracked ice, put in a few spoonfuls cream and let it get firm; put over this some of the soaked lady fingers and macaroons and over them some preserved pineapple or cherries; over this put cream, fruit and cake; continue in this way until all is used; let the last layer be cream; close the form and pack it in cracked ice, where it should remain 2 hours; when ready to serve turn the pudding onto a round dish, fill the opening in center with whipped cream flavored with vanilla and garnish the edge of dish with preserved fruit.

211.Chocolate Pudding à la Hollandaise.— Boil ¼ pound Baker's grated chocolate in ½ pint water, add ½ pint sugar and 1 teaspoonful extract of vanilla; when cold add 1½ ounces gelatine which has been soaked in ½ pint cold water and dissolved in ½ pint hot water; set the chocolate mixture into cracked ice and stir till it begins to thicken; then add 1 pint whipped cream; if not sweet enough add a little more sugar; set a tin pudding form with a tube in the center into cracked ice, pour in some clear fruit or wine jelly (see Jelly) and let it get firm; decorate the bottom with blanched almonds; take pieces of almonds up with a larding needle, dip them into jelly and lay in a pointed border close to the edge; pour 51 over a little more jelly; in the meantime soak 20 vanilla wafers and macaroons in sweet cream; when the jelly in form is firm put in a layer of wafers and macaroons; put over this a layer of the chocolate cream; as soon as the cream is firm put in the remaining wafers and macaroons and lastly the remaining cream; let the pudding remain on ice for about 3 hours; when ready to serve dip the form into hot water, turn the pudding onto a round dish and lay a border of whipped cream flavored with extract of vanilla around it; fill the opening in center with whipped cream.

212.**Pudding à la Reine.**— Set a border form into cracked ice and pour in to the depth of about ½ inch some white wine jelly; when the jelly is firm put in some fruit, such as strawberries, cherries, plums or peaches, and pour over a few spoonfuls jelly; after the lapse of 5 minutes pour in more jelly; when firm put in another layer of fruit and then fill the form with jelly; let it remain on the ice till ready to serve; pare and cut into slices 12 large, ripe peaches, sprinkle thickly with sugar and let them stand 1 hour; press them with the juice through a sieve, add 1½ ounces gelatine dissolved in water, set on ice and stir till it begins to thicken; then add 1 pint whipped cream, 1 glass sherry wine and a few lady fingers broken into pieces; fill the cream into a highly pointed form and set it into cracked ice for 2 hours; when ready to serve turn out the jelly from the border form onto a round dish; then turn out the cream from the highly pointed form; place the latter in the center of the jelly border and serve.

213.**Pudding à l'Allemande.**— Boil 1½ pints milk with 4 table-spoonfuls sugar, ¼ teaspoonful salt and the thin peel of 1 lemon; mix 1 cup flour with 1 cup milk to a smooth paste and add it to the boiling milk, stirring constantly; boil a few minutes, remove from the fire, add the beaten yolks of 6 eggs and stir until nearly cold; then add the whites of the 6 eggs, beaten to a stiff froth; rinse a jelly form with cold water, sprinkle with sugar, pour in the mixture and place on ice for 3 or 4 hours; in serving turn the pudding onto a dish, garnish with strawberries and serve with the following sauce:—Boil 2 teaspoonfuls cornstarch in 1½ cups water, sweeten with sugar, 52 remove from the fire, add the juice of 1 lemon, ½ pint strawberry juice, 1 glass Rhine wine and serve when cold with the pudding.

214.**Strawberry Pudding.**— Place a round tin form into cracked ice and pour in some orange jelly; when firm lay the form over on its side, pour in more jelly, turn the form around and pour in more jelly; continue in this way until the whole inside of form is glazed with the jelly; mix 1 pint bruised strawberries with 1 pint sugar syrup flavored with vanilla and add 1½ ounces dissolved gelatine; put this on ice and stir till it begins to thicken; pour some Madeira wine over some lady fingers and let them soak about 10 minutes; put a layer of the strawberry pureé in the form, over this some lady fingers and continue with cream and cake in alternate layers till all is used; let the form remain on ice for 2 hours; then turn the pudding onto a dish, garnish with chopped orange jelly and nice, large strawberries which have been dipped into the jelly and serve with strawberry syrup.

215.**Imperial Pudding.**— Place a cream form into cracked ice and pour in some white wine jelly colored to a delicate pink with cochineal; when the jelly is firm decorate the bottom with preserved pineapple cut into the shape of dice and blanched almonds cut into strips; pour over a few spoonfuls jelly and let it remain till firm; place a saucepan with 1 pint cream, the yolks of 6 eggs and 5 tablespoonfuls sugar over the fire and stir until nearly boiling; when cold add 1½ ounces dissolved gelatine and ½ pint best arrac; soak 1 dozen vanilla wafers and the same quantity of macaroons in sugar syrup mixed with champagne and arrac for 10 minutes; stir the cream on ice until it begins to thicken; then add 1 pint whipped cream and lastly ½ pint champagne; fill the cream in alternate layers with wafers and macaroons in the form; let the pudding remain for 2 hours on ice; pour into tartlet forms some orange jelly with small dice of pineapple; in serving dip the form into hot water and turn the pudding onto a round dish; also turn out the jelly from the small moulds and lay them around the dish.

53

216.Suédoise of Apples. — Pare 1 dozen large apples, bore pegs therefrom with an apple corer and lay them in water with lemon juice; prepare 1 dozen large Bartlet pears the same way; boil the apple pegs in sugar syrup with lemon juice, to keep them white, and boil the pears in sugar syrup with cochineal; care must be taken not to boil them too long, so that they will not fall apart; transfer them to a dish and set aside to cool; wash the apple and pear peels, also the cores; put them in a saucepan with sufficient water to cover and boil till done; strain through a jelly bag; measure the liquor and take for 1 quart 1½ ounces gelatine, the thin peel and juice of 1 lemon, 1 cup sugar and the whites of 2 eggs; soak the gelatine in a little cold water 15 minutes; put the liquor with lemon, sugar and well beaten whites over the fire; when hot add the gelatine, stir constantly and boil 5 minutes; remove to side of stove, add ½ pint white wine and strain through a jelly bag; place a plain form with tube in the center into cracked ice and pour a few spoonfuls jelly in the bottom of it; when firm lay the form over on its side, pour in more jelly, keep turning and add by degrees more jelly; continue this process until the jelly has formed a complete lining inside of form; lay the pegs of apples and pears in slanting rows onto a napkin and cut them all the same length; then take each one separately onto a larding or knitting needle and dip into cold jelly; first lay a row of red on the side of form, then a row of white in an opposite direction; continue until the form is covered, pour over some thick jelly and when firm fill the inside with apple bavarois made as follows: — Prepare 1 pint apple sauce, press it through a sieve, add 1 teaspoonful vanilla and sweeten to taste; soak 16 sheets gelatine in cold water for 10 minutes, press out, put in a bowl and pour ½ cup boiling water over it; stir until dissolved, add to the apples and stir until it begins to thicken; then mix in 1 pint whipped cream or the beaten whites of 6 eggs; fill this into the form, cover and let it remain on ice till firm; in serving dip the form into hot water, dry it quickly, turn the suédoise onto a round dish and garnish with fruit; the apple may be bored out into rounds like marbles and boiled the same way — half red and half white; they are then laid in rows on the side of form over one another, alternately with white and red 54 till the form is lined with them; then place a small form inside, pour sufficient jelly around to cover the fruit and fill up the space between the inside form and fruit; let it remain on ice till firm; then pour in the inside form some hot water, draw it out and fill the inside with any kind of frozen cream; serve at once.

217.**Suédoise of Pears.**— Pare and quarter 12 large Bartlet or duchess pears and cut each quarter lengthwise into 4 slices; boil half the slices in sugar syrup with lemon juice and the other half in sugar syrup with cochineal; lay them on a napkin to dry; pour a little wine jelly into a plain form and lay on the bottom some of the slices in the shape of a star; when firm turn the form over on its side and lay in first a row of white slices, then a row of red; dip each piece into cold wine jelly before laying it in the form; continue in this way until the sides of form are covered; then pour in a few spoonfuls jelly and keep turning the form, in order that the jelly may be evenly distributed over the fruit; pare and cut into small pieces ½ dozen large pears, put them over the fire with a little water and boil till soft; press them through a sieve and set aside to cool; boil 1 cup sugar in 1 cup water with the juice of ½ lemon for a few minutes; soak 16 sheets gelatine in cold water 5 minutes, press out, add it to the sugar and boil a little longer; remove from fire, mix with the pear pureé and stir till it begins to thicken; whip 1 pint sweet cream to a froth, add 2 tablespoonfuls powdered sugar and 1 teaspoonful vanilla extract; add it to the above pear mixture and fill into the form; place it on ice for 2 hours; when ready to serve turn the suédoise onto a round dish and garnish with croutons of wine jelly.

218.**Timbale de Pêche à la Condé.**— Line a deep round form with rich pie crust, lay buttered paper over it, fill the form with dry peas and put in oven to bake; when baked take it from the oven, remove the peas, return form to oven and let the crust dry for a few minutes; place 1 cup rice with cold water over the fire and boil a few minutes; drain in colander, rinse with water and boil in milk till soft and thick; add ½ cup sugar, ½ tablespoonful butter and set it in a warm place; pare and cut into halves 1 dozen large, 55 ripe peaches and boil a few minutes in sugar syrup; draw them to side of stove to keep warm; also have the form with crust (or timbale) setting in a warm place; mix ½ cup whipped cream with the rice and fill it alternately with the peaches in the form inside of timbale; let the last layer be rice; put a round dish over the form and turn the timbale onto it; cut a round hole in the center, put in a few peaches and pour the peach syrup all over the timbale. Timbale of cherries, apricots, pineapples, pears and apples are made in the same manner.

219.**Timbale de Riz à la Napolitaine.**— Put ¾ pound parboiled rice with 1 quart milk, ½ teaspoonful salt, 1½ tablespoonfuls butter and a little vanilla over the fire and boil till rice is tender; when done add some seedless raisins, currants and fine citron (1 cupful in all) and set aside to cool; stir 4 tablespoonfuls sugar with 1 whole egg and the yolks of 4 to a cream; add 2 tablespoonfuls Madeira wine and mix it with the rice; line a deep round form with thin neapolitan paste, fill it with the rice, put on a cover of the same dough and bake 1 hour; when baked turn the timbale onto a dish, pour over it a fruit sauce mixed with Madeira wine and send some in a saucere to table with it.

220.**Pear Timbale.**— Pare, quarter and stew 1 dozen Bartlet pears with 1 bottle claret, 1 cup sugar, asmall piece of cinnamon and ½ cup seedless raisins; when done pour them on a sieve to drain and cool; line a buttered, deep round form or tin pan with about 1 inch of biroche dough (see Biroche), fill with the pears, put on a cover of the same dough and let it stand in a warm place for ½ hour; then bake in a medium hot oven; when baked turn the timbale onto a round dish, pour some of the pear syrup over and serve the rest in a saucere with it.

221.**Timbale à la Sicilienne.**— Butter a deep round form, line it with neapolitan paste, cover the latter with buttered paper, fill the form with dry peas and bake in a hot oven; when done and cold remove peas and paper, take the timbale from the form, brush over the inside and outside with peach or apricot marmalade and decorate it aroundA and on top with blanched half almonds and currants; take 56 a form 1 inch wider than the one above, place it into cracked ice and pour in, to the depth of about ¾ inch, some clear lemon jelly; as soon as cold place timbale into the form and fill space between the timbale and form with lukewarm lemon jelly; let it remain on ice till needed; when ready to serve fill the timbale with peach, pineapple or strawberry plombiere or any kind of frozen cream; dip the form into warm water, dry quickly, turn it onto a round dish and decorate with sugared orange quarters.

222.**Timbale of Mixed Fruit.**— Take some preserved peaches, pineapples, cherries and pears and put them on a sieve to drain; then put them in a dish with ½ cup currants or apple jelly and 1 teaspoonful vanilla sugar; mix all together and fill it into a form lined with biroche dough; cover with the same dough and finish same as Pear Timbale.

223.**Chocolate Plombiere.**— Dissolve ½ pound grated chocolate in ½ cup water, add the yolks of 8 eggs, 1 pint cream, 1 teaspoonful vanilla extract, 6 tablespoonfuls sugar and stir this over the fire till nearly boiling; strain through a hair sieve and when cold put it in a freezer; finish the same as Orange Plombiere.

224.**Orange Plombiere.**— Strain the juice of 6 oranges and rub the skin from 2 with loaf sugar; dissolve ¾ pound sugar in 1 cup cold water and mix it with the orange juice and orange sugar; put into a freezer and turn and work it till it thickens; then add 1 pint whipped cream and work it for 10 minutes longer; then fill the mixture into a form, cover tightly and paste a strip of buttered paper around the edge of cover; then pack the form into cracked ice and salt; lay plenty of ice on top and let it remain from 1 to 2 hours.

225.**Rum Plombiere.**— Place a saucepan with 1 pint cream, the yolks of 10 eggs and 1 cup sugar over the fire and stir till nearly boiling; remove from the fire and set aside to cool; cut 6 ounces candied orange peel into small dice and boil them for a few minutes in a little water; drain on a sieve, add them to the cream and put the mixture into a freezer; let it freeze till it begins to 57 thicken; then add ½ cup best rum and 1 pint whipped cream; fill the mixture into a form, paste a strip of buttered paper around the edge of cover and pack in ice and salt for 2 hours.

226.**Plombiere of Maraschino Curaçoa** is made the same way.

227.**Pistache Plombiere.**— Pound ¼ pound blanched almonds and ¼ pound blanched pistachio nuts with a little cream to a paste; place a saucepan with the paste, 1½ pints cream, 1 cup sugar, 1 teaspoonful vanilla and the yolks of 8 eggs over the fire and stir until nearly boiling; remove cream from fire, set it in cold water and stir till cold; add a little spinach color and strain through a hair sieve; then finish the same as Strawberry Plombiere.

228.**Plombiere** aux Café.— Pour 1½ pints boiling cream over 3 tablespoonfuls freshly ground coffee and let it stand well covered for 10 minutes; strain through a napkin; put the coffee cream in a saucepan over the fire with 1 cup sugar and the yolks of 8 eggs and stir till nearly boiling; remove the cream from fire, set saucepan in cold water and stir till cold; then finish the same as Strawberry Plombiere.

229.**Tea Plombiere.**— Pour 1½ pints boiling cream over 1 ounce tea and let it stand 5 minutes; strain and finish the same as Coffee Plombiere.

230.**Peach Plombiere.**— Pare, quarter and press through a sieve 15 large, ripe peaches; dissolve ¾ pound sugar in 1 cup water and add it to the peach pureé; put this mixture into the freezer and finish the same as Strawberry Plombiere.

231.**Vanilla Plombiere.**— Place a saucepan with 3 cups milk over the fire, add the yolks of 8 eggs, 1 cup sugar and 2 teaspoonfuls vanilla extract and stir till nearly boiling; remove from the fire, set the saucepan into cold water and stir till cool; then put into a freezer and let it freeze till it begins to thicken; then add 1 pint whipped cream and finish the same as Strawberry Plombiere.

58

232.**Strawberry Plombiere.**— Wash 1 quart strawberries and press them through a sieve; dissolve ¾ pound sugar in ¾ cup water and add this syrup to the strawberry pureé; 2 hours before serving pour it into a freezer and turn it about 20 minutes, or till it begins to thicken; then mix with 1 pint whipped cream and let it remain a little while longer in the freezer; fill into a form, cover tightly, paste a strip of buttered paper around the edge of cover and pack in ice and salt for 2 hours; in serving dip form into hot water, quickly wipe it dry, turn the plombiere onto a round dish and garnish with fancy cake.

233.**Pineapple Plombiere.**— Pare and cut into small dice 1 ripe pineapple, put them into a dish and pour 1 pint cold sugar syrup over it; let it stand 4 hours; 2 hours before serving put the fruit into a freezer and freeze till it begins to thicken; then add 1 pint whipped cream and finish the same as in foregoing recipe. NOTE.— This plombiere may also be served in a glass dish directly from the freezer; it must then, of course, be worked until firm. If preserved fruit is used less sugar must be taken, and color and taste should be freshened up with lemon juice and a few drops of cochineal. Plombiere of raspberries, currants or cherries is made in a similar manner.

FROZEN PUDDINGS.

234.**Frozen Strawberry Pudding.**— Whip 1 quart rich, sweet cream until thick, add 2 cups powdered sugar and lastly stir 1 quart mashed strawberries through the cream; fill this into a pudding form with a tube in the center, cover tightly and put a strip of buttered paper around the edge of cover, so that no water can enter; have ready a large pail or a butter tub, put some cracked ice on the bottom, sprinkle over some rock salt, set onto this the form, fill up the sides with cracked ice and sprinkle salt between; cover the top of form with ice, the whole with a piece of carpet or a cloth and set in a cool place for 4 hours; when ready to serve lift 59 from the ice, remove the paper, wipe off the form, dip it in hot water, turn the pudding onto a dish and serve at once.

235.**Rich Ice Cream Pudding.**— Beat the yolks of 9 eggs with ¾ pound sugar to a cream and add 1 quart whipped cream; fill this into a tin pudding form with a tube in the center, paste over the edge of cover a strip of buttered paper and bury in cracked ice and rock salt for 4 hours, the same as Strawberry Pudding.

236.**Pudding à la Pückler Muskau.**— Stir into 1 quart whipped cream 6 tablespoonfuls sugar and 6 ounces finely pounded macaroons; fill the cream into a form and bury it in ice and rock salt for 4 hours, the same as Strawberry Pudding.

237.**Ice Pudding à la Prince Pückler.**— Whip 1 quart cream till stiff and divide it into 3 parts; boil 6 ounces grated chocolate in ½ pint water with ½ cup sugar smooth and thick; remove the chocolate from the fire and when cold mix with it ⅓ of the whipped cream; mix 1 pint bruised raspberries with another ⅓ of the whipped cream and add sufficient sugar to sweeten (or take raspberry jelly); add to the last ⅓ of whipped cream 5 tablespoonfuls sugar and 1 teaspoonful vanilla extract; place a form into cracked ice, fill in the cream in finger thick layers alternately—first the chocolate, then the raspberry, then the white; continue until all is used; cover the form tightly, paste around the edge of cover a strip of buttered paper and bury the whole form in rock salt and ice for 4 hours; if the ice melts more must be put around the form and some of the water drawn off; when ready to serve dip the form into hot water, turn the pudding onto a dish and serve at once.

238.**Frozen Chocolate Pudding.**— Boil 6 ounces grated chocolate in ½ pint water with 4 tablespoonfuls sugar until thick and smooth; when cold mix it with 1½ pints whipped cream; if not sweet enough add more sugar; fill this into a tin pudding form, paste a strip of buttered paper around the edge of cover and bury the form in cracked ice and rock salt for 4 hours the same way as Strawberry Pudding.

60

239.**Bombe à la Altenberg.**— Boil 1½ cups sugar with 1 cup water 10 minutes; remove and when cold add the yolks of 6 eggs: stir this over the fire till nearly boiling; when cold mix it with 1 pint whipped cream and 1 teaspoonful vanilla extract; fill this into a form, cover tightly and place into cracked ice; boil 6 ounces chocolate in 1 cup water with ½ cup sugar and 1 teaspoonful vanilla until smooth; put this into another form, also standing in ice and rock salt; when it begins to freeze spread the chocolate evenly around the inside of form, so as to form a complete lining; then cover the form and let it remain in ice until hard; next fill in the above vanilla cream, cover tightly, paste a strip of buttered paper around the edge of cover and bury the form in plenty of ice and rock salt for 4 hours; in serving dip the form in hot water, quickly turn the bombe out onto a round dish, decorate with kisses and serve at once.

240.**Bombe à la Parisienne.**— Press 1 quart ripe strawberries through a sieve, add 1 pound sugar dissolved in ½ pint cold water and a little Rhine wine; pack a plain ice cream form into cracked ice and salt, pour in the strawberries and let freeze till it begins to thicken; then spread the half frozen strawberry ice onto the sides and bottom of form so that it forms a complete lining inside; cover the form and let it remain in ice till hard; in the meantime have a pineapple cream prepared as follows: —

241.Pineapple Cream for Bombe à la Parisienne.— Place a saucepan with the yolks of 6 eggs and 1 pint pineapple syrup over the fire and stir until nearly boiling; remove from the fire and when cold add 1 pint whipped cream; fill this inside of the strawberry ice, cover the form tightly, paste a strip of buttered paper around the edge of cover and bury in ice and salt for 3 hours; when ready to serve take out the form, rinse off with cold water, remove the paper, dip the form quickly into hot water and turn the bombe onto a handsome dish; garnish with fruit, French candies or fancy cakes and serve at once. NOTE.—The strawberry ice may be first frozen in a freezer and then put into the form.

61

242.Ice Pudding (with Pumpernickel).— Cut 6 ounces pumpernickel into slices and dry them in the oven; roll them fine with a rolling pin and sift the crumbs through a coarse sieve; mix them with 1 quart whipped cream, add 1 teaspoonful extract of vanilla and 1 cup sugar; fill the cream into a tin form with a tube in the center, cover tightly, paste a strip of buttered paper around the edge of cover and bury in cracked ice and rock salt for 4 hours; when ready to serve dip the form into hot water, turn the pudding onto a round dish and serve at once.

243.Ice Pudding (with Almonds).— Stir the yolks of 8 eggs with ¾ pound sugar to a cream, add 1 quart whipped cream, ½ pound ground almonds and finish the same as Strawberry Pudding.

244.Frozen Pudding à la Montmorency.— Mix 2 tablespoonfuls sugar with 1 cup finely chopped sweet almonds and 10 bitter ones; put this into a tin pan and roast in the oven to a light brown, stirring often; place a saucepan with the yolks of 6 eggs, 1 cup sugar, 1 pint cream or milk and the roasted almonds over the fire and stir constantly until nearly boiling; then strain through a sieve; when cold add 2 tablespoonfuls caramel (see Boiling Sugar) and orange blossom water; put this into an ice cream freezer and work till it begins to thicken; then add ¾ pint whipped cream, ½ cup finely chopped pistachio nuts and 3 ounces finely powdered macaroons; continue working the freezer till the cream is frozen hard; place a cream form in ice and salt, pour some cherry syrup around the sides and bottom, sprinkle with pistachio fillets and some preserved red cherries; then fill in the cream with some of the cherries laid between, put on the cover, paste a strip of buttered paper around its edge and completely bury the form in ice and rock salt for 1 hour; when ready to serve lift from the ice, rinse off with cold water, remove the paper, wipe the form dry and quickly dip it into hot water; have ready a handsome dish with a folded napkin; turn the pudding onto the dish, garnish with small fancy cakes and serve with whipped cream flavored with vanilla or maraschino.

62

245.Pudding Glacé à la Metternich. — Pound 3 ounces blanched almonds and 3 ounces blanched pistachio nuts to a paste; stir over the fire the yolks of 6 eggs, 6 tablespoonfuls sugar, 1 pint cream and a little vanilla till nearly boiling; add the almonds and pistachio paste and set aside to cool; then strain through a sieve; soak 6 ounces seedless raisins, 3 ounces finely cut preserved orange peel and a little finely cut preserved apricots in maraschino and cut a small sponge cake into slices; 4 hours before serving place a high form into cracked ice and salt, put in a layer of cream and over this some fruit and cake; continue with cream, fruit and cake alternately till all is used; cover the form, paste a piece of buttered paper around the edge of cover and completely bury in plenty of cracked ice and salt; when ready to serve rinse the form off with cold water, remove the paper, quickly dip the form into hot water, turn the pudding onto a dish and garnish with fruit and fancy cake; serve with pistachio sauce made as follows: — Stir the yolks of 4 eggs with 1 pint sweet cream and 2 tablespoonfuls sugar over the fire till nearly boiling; remove from the fire, add 2 ounces finely powdered pistachio nuts and serve when cold with the pudding.

246.A la Duchesse de Berry. — Press 1 quart strawberries through a sieve; dissolve ¾ pound sugar in 1 cup cold water and add it to the strawberries; set a form into cracked ice and salt for 20 minutes, put in the strawberries and freeze it until thick; then spread the strawberry ice around the sides and bottom of a high ice form and let it stand in ice till hard; in the meantime prepare a cream for the inside; mix 1 pint cream with 6 tablespoonfuls sugar, the yolks of 6 eggs and 1 cup preserved pineapple cut into small pieces; stir this over the fire till nearly boiling; remove it and when cold put into a freezer; freeze until it begins to thicken; then add 1 pint whipped cream and freeze it for a few minutes longer; then fill it into the strawberry form, cover tightly, paste a strip of buttered paper around the edge of cover and bury in plenty of ice and rock salt for 1 hour; in serving take out of ice, rinse off with cold water, remove the paper, wipe the form dry, quickly dip it into hot water, turn the pudding onto a handsome dish and serve at once.

63

247.**Pudding Glacé à la Allemande.**— Put 2 dozen lady fingers on a long plate and pour over them some Madeira wine or maraschino; set a plain form without a tube in ice and rock salt; stir 1 pint cream with the yolks of 6 eggs, 6 tablespoonfuls sugar and 1 teaspoonful vanilla extract over the fire till nearly boiling; when cold put in a freezer and freeze till it begins to thicken; then add 1 pint whipped cream and freeze for a few minutes longer; then put a layer of this cream into the plain form, standing in ice, put over this a layer of lady fingers and a few spoonfuls apricot marmalade or fruit jelly, then a layer of cream again; continue this way until all is used; let the last layer be cream; put on the cover, paste a piece of buttered paper around the edge of it and bury the form completely in ice and rock salt; let it remain 1 hour; then serve, garnished with fancy cake. NOTE.—If a form is not handy a 3-quart tin kettle will do.

248.**Frozen Pudding à la Richelieu.**— Boil ¼ pound rice in water till done and pour it onto a sieve to drain; pound ¼ pound blanched almonds or pistachio nuts with a little cream to a paste; remove the shells and brown skin from ¼ pound large chestnuts and boil them in milk till soft; then strain them through a sieve and mix rice and nuts together; boil ½ cup sugar with ½ cup water for 10 minutes, add 1 teaspoonful vanilla extract, mix (hot) with the above mixture and let it stand for an hour; put in a porcelain-lined saucepan 1 pint cream, the yolks of 6 eggs, 6 tablespoonfuls sugar and 1 teaspoonful vanilla extract; stir this over the fire till nearly boiling; remove it and set aside to cool; spread 20 vanilla wafers on one side with apricot marmalade and put 2 and 2 together; dip them into sherry wine and set aside; also cut some stewed pineapple into dice; set a form into cracked ice and salt and put in a few spoonfuls cream; lay over the cream a layer of wafers, rice and pineapple; then cream again; continue until all is used; put on the cover, paste a strip of buttered paper around its edge and bury the form completely in ice and rock salt from 3 to 4 hours; when ready to serve turn the pudding onto a dish with a folded napkin underneath and send cold pineapple or pistachio sauce to table with it.

64

249.**Frozen Chestnut Pudding.** — Place a saucepan with ½ pound large chestnuts over the fire, cover with water and boil a few minutes; drain the nuts in colander, remove the outside shell and the inside skin and boil in milk till soft; press them through a sieve and add the yolks of 6 eggs to the pureé and 1 pint sweet cream, ½ pound sugar and 1 teaspoonful extract of vanilla; stir this over the fire till nearly boiling; strain through a fine sieve; boil for 15 minutes 2 ounces well washed currants, the same of seedless raisins and finely cut citron and a little orange peel in water; drain on a sieve and let them lay for 2 hours in Madeira wine; put a piece of ice (large enough to cover the bottom) in a strong pail or butter tub and sprinkle a handful of rock salt over it; put onto this an ice cream freezer and fill up the sides with cracked ice and salt; put in the chestnut cream and work till it begins to thicken; then pour in not quite 1 pint whipped cream and work until it is frozen quite hard; then add the fruit with the wine; let it freeze a little longer; transfer the cream to a pudding form with a tube in the center (or an ice form), put on the cover, paste a strip of buttered paper around its edge and bury the form in ice and salt for 1 hour; when ready to serve rinse off the form with cold water, remove the paper and wipe dry; then dip it quickly into hot water and turn the pudding onto a dish; garnish with fancy cake and serve with whipped cream.

250.**Frozen Apple Pudding.** — Pare and core 8 nice greening apples, cut them into quarters and stew with ½ cup water till tender; boil 1 cup sugar with 1 cup water for 5 minutes and add it to the apples with 1 teaspoonful vanilla extract and ½ cup apricot marmalade; press the whole through a sieve; when cold put it into an ice cream freezer and work till it begins to get thick; then add 1 pint whipped cream, 3 ounces currants and the same of seedless raisins and finely cut citron; the last 3 ingredients should be boiled for 20 minutes in a little water and laid for ½ hour in vanilla syrup; let the whole freeze until hard; fill the cream into a form, put on the cover, paste a strip of buttered paper around the edge of the latter and bury in salt and ice for 1 hour; serve with whipped cream and garnish with fancy cakes.

65

251.**Mousse of Pineapple.**— Line a plain form with white paper; see that there are no creases in the paper; lay it in even and smooth; set the form into cracked ice until following mixture is prepared:— Pare and cut into slices 1 ripe pineapple; dissolve 1 pound sugar in 1 pint water and put it over the fire to boil; add the pineapple slices and boil 20 minutes; transfer them to a sieve to drain; when cold cut some of the slices into halves and lay them inside on the side of form; cut the remaining slices of pineapple into dice and set them cold; place a saucepan with 1½ cups pineapple syrup and the yolks of 9 eggs over the fire and stir till nearly boiling; remove from fire, add 1 cup pineapple dice and stir till cold; then mix it with 1 pint whipped cream; fill this into the form, put on the cover and paste a strip of buttered paper around its edge; then pack the form into cracked ice and salt so that it is completely buried and let it remain 4 hours; when ready to serve dip the form into hot water, dry quickly, turn the mousse onto a dish and garnish with fancy cakes.

252.**Mousse à la Vanille.**— Dissolve 1 cup sugar in 1½ cups water, add the yolks of 6 eggs and stir over the fire till nearly boiling; remove quickly and stir till cold; then add 1 pint whipped cream, 2 teaspoonfuls vanilla extract and finish the same as pineapple in foregoing recipe.

253.**Mousse à l'Orange.**— Dissolve 1 cup sugar in 1 cup water and boil a few minutes with the juice of 1 lemon; remove the syrup from the fire, put in the thin peel of 2 oranges and let them lay for a few minutes; then remove; rub off the skin from 6 oranges with loaf sugar and add the orange sugar to the sugar syrup with the juice of 6 oranges and the yolks of 9 eggs; beat this with an egg beater till nearly boiling; remove quickly, set it in cold water and continue beating till cold; then add 1 pint whipped cream and finish the same as Pineapple Mousse.

254.**Mousse au Chocolat.** — Dissolve 3 ounces grated chocolate in ½ cup water and boil for a few minutes; strain through a sieve and set aside; put in a saucepan the yolks of 6 eggs and 1 cup 66 sugar syrup and stir over the fire till it begins to thicken; remove it quickly, set saucepan in cold water, add the chocolate and stir till cold; then mix it with 1 pint whipped cream and finish the same as Pineapple Mousse.

255.**Mousse au Maraskino.** — Stir the yolks of 6 eggs with ¾ cup sugar and 1 cup water over the fire to a cream; remove it from the fire and stir till cold; add ½ cup maraschino and 1 pint whipped cream and finish the same as Pineapple Mousse. Rum may be substituted for maraschino.

PAINS.

256.**Pain of Strawberries.** — Put 1 quart ripe strawberries into a colander, rinse off with cold water and press them through a sieve; soak 2 ounces gelatine in ½ pint cold water for 15 minutes, add ½ pint boiling water and stir over the fire till gelatine is dissolved; set aside to cool; then dissolve ¾ pound sugar in 1 pint cold water, put it over the fire with the juice of 1 lemon and boil 5 minutes; when cold add it with the gelatine to the strawberries; also add ½ cup white wine and a little cochineal; put the pain on ice till it begins to thicken; then fill it into a form with a tube in the center, cover and place for 2 or 3 hours on ice. Pains of raspberries or currants are made the same way, using no lemon.

257.**Pain d'Ananas.** — Take a jar of preserved pineapples, cut them into small dice, add ½ pint white wine and a little more sugar if necessary; add the juice of 1 lemon and 2 ounces gelatine dissolved in 1 pint water; place this on ice and stir it now and then; as soon as it begins to thicken put into a form, which set on ice for 2 or 3 hours; then serve.

258.**Pain d'Ananas à la Parisienne.** — Chose a large, ripe pineapple and pare and grate it; add 1½ cups sugar and stir until dissolved; press the pineapple through a sieve and add the juice of 4 oranges and 2 ounces gelatine dissolved in 1 pint water; place a plain form into cracked ice and pour in a few spoonfuls orange jelly; 67 when this is hard lay the form over on its side, pour in more jelly and keep turning slowly, so that the jelly may get all over the sides and form a lining; next have some pistachio nuts or blanched almonds cut into strips and sprinkle them over the sides and bottom of form; set the pineapple mixture on ice and stir until it begins to thicken; then fill into the form, cover and let it remain for 2 or 3 hours in ice; it is then ready to serve; chop the trimmings of the pineapple fine, pour over some cold sugar syrup and let it stand 2 hours; strain, add a little dissolved gelatine and pour over the pain when sent to table.

259.**Pain de Peches.**— Take 20 large, ripe peaches and pare and quarter them; then press them through a sieve; add to this 1 pound sugar dissolved in 1 pint cold water and 2 ounces dissolved gelatine; crack the stones, remove the pits, scald in boiling water and free them from their brown skin; cut the pits in half and boil them in a little sugar syrup; add to the peach mixture ½ cup white wine and fill it into a tin form with a tube in the center; place the form on ice and let it remain till its contents begin to thicken; then stir in the peach pits and let it remain on ice 2 hours longer. Pains of apricots, cherries or plums are made the same way.

260.**Pain à la Victoria.**— Press 1 pint ripe raspberries through a sieve and mix it with 1 ounce gelatine dissolved in ½ pint water; put ¾ cup sugar into ½ pint cold water and stir until dissolved; then add it to the raspberries with a glass of white wine; place this on ice till it begins to thicken; prepare 1 quart almond blanc-mange (see Blanc-Mange); set a plain form into cracked ice and put in a layer of raspberries about an inch in thickness; let this get hard; then put in a layer of blanc-mange; after this is firm again put in raspberries, then blanc-mange; continue till all is used; let it remain on ice for 2 hours; when ready to serve turn the pain onto a round dish and garnish with fruit.

261.**Pain de Peches à la Richelieu.**— Prepare a pain the same as Pain de Peches and also 1 pint almond blanc-mange; set a plain form with a tube in the center into cracked ice and put in by 68 degrees the blanc-mange; put it ½ inch in thickness all around on the sides and bottom of form, so that it forms a complete lining inside;

then fill in the pain of peaches and let it remain on ice for 2 hours: in serving dip the form into hot water, wipe dry and turn its contents into a glass dish.

262.**Pain de Peches à la Condé.**— Pare and cut into halves 1½ dozen large, ripe peaches and boil them with their blanched pits in sugar syrup for about 10 minutes; transfer the peaches to a dish or long tin pan, wipe dry and lay them with the hollow side up; put half a pit in the center of each and pour a spoonful of jelly over each piece (the jelly should be previously stirred on ice till it begins to thicken); next set a plain form into cracked ice, pour in some plain fruit or wine jelly and keep turning the form until the inside is lined with the jelly; cover the bottom with peaches; lay them so that the pits are to the outside; then lay the remaining peaches in rows on the side of form, pour over some jelly and when firm fill up the form with bavarois aux apricots, which is prepared as follows:— Pare and cut into pieces 1½ dozen ripe apricots, lay in a dish, sprinkle over 1 cup sugar and let them stand for 2 hours; then press them through a sieve; mix the pureé with 1 teaspoonful extract of vanilla and 1½ ounces dissolved gelatine; put this on ice and stir till it begins to thicken; then carefully stir 1 pint whipped cream through it; fill the bavarois into form at once and let it remain on ice for 2 or 3 hours.

263.**Pain de Pommes à la Condé.**— Choose 15 large pippin or greening apples and pare, quarter and stew them with a little water; press them through a sieve, add 1 cup sugar and when cold mix it with 2 ounces gelatine dissolved in 1 pint water and 1 pint almond milk (prepared the same as for Blanc-Mange); place this on ice and stir till it begins to thicken; fill it into a form which has already been set into cracked ice and let it remain for 2 or 3 hours; boil the cores and peels of apples till soft; strain through a bag, return the liquor to saucepan and boil 10 minutes; then add to ½ pint of juice 1 cup sugar and boil for a few minutes; pour the syrup into a dish and set aside to cool; in serving dip the form into hot 69 water, wipe it dry and turn the pain onto a round dish, or into a large glass dish, and pour the apple syrup over it. This pain may be made of Bartlet pears in the same manner.

ICE CREAMS.

264.**Directions for Making Ice Cream.** — The implements needed are a freezer, rock salt and finely cracked ice. Ice cream freezers can be bought at any hardware store. They consist of a large wooden pail with a faucet on the side near the bottom and a freezer with a paddle inside. The cracking of the ice is best accomplished by putting it into a coarse sack and pounding it fine with a hammer or mallet. Place the freezer into the pail, put in the paddle and cover the freezer tightly. Fill the space between the pail and freezer with fine cracked ice to ⅓ its height, sprinkle over 2 handfuls salt and pack down the ice with a piece of wood, so that it may be firm all around the freezer; continue with layers of ice, salt and the packing down till the ice reaches to the edge of cover; next pour into the freezer the mixture that is to be frozen; but care should be taken not to put in too much, for the cream needs plenty of room in order to become light and smooth; cover the freezer and let it stand for 5 minutes; then commence to turn; after 10 minutes' turning remove the cover from freezer and cut the frozen cream with a long bladed knife from the sides of can; repeat this every 10 minutes until the cream is frozen hard; then remove the paddle, even off the cream in the freezer, cover and let it stand for 10 minutes; do not draw off the water from pail until it stands above the ice and the freezer has lost its firm hold; after drawing off the water fill the space up again with cracked ice and salt; when the 10 minutes have elapsed fill the frozen cream into an ice form, cover tightly and paste a strip of buttered paper around the edge of cover; then pack the form into cracked ice and salt for 1 or 2 hours; when ready to serve take the form from the ice, rinse it off with cold water, remove the 70 paper and wipe the form dry; then dip it quickly into hot water, take off the cover, turn the cream onto a dish and serve at once.

265.**Ice Cream (large quantity).** — 14 quarts sweet cream, 6 quarts milk, 7 pounds sugar, 30 eggs and ¼ pound gelatine; soak gelatine for 10 minutes in a little of the milk; put the remaining milk over the fire and boil; then add the soaked gelatine and stir and boil till it is dissolved; set aside to cool a little; beat eggs and sugar to a cream and add by degrees the milk, stirring constantly; return to fire and let it get boiling hot; but do not allow it to boil, otherwise it will curdle; remove from fire, pass it through a sieve and set aside to cool, stirring it occasionally; beat the cream until quite thick, gradually add the cold custard and continue beating for a little while longer; then put it in a freezer and freeze as directed.

266.**Fine Vanilla Ice Cream.**— Beat the yolks of 8 eggs to a cream and add gradually 1 quart sweet cream which has previously been boiled and cooled; add ¾ pound sugar, 2 teaspoonfuls vanilla extract and stir the whole over the fire until nearly boiling; then remove from fire and when cold strain it through a sieve and freeze as directed.

267.**Custard Ice Cream.**— Put 5 eggs in a saucepan and beat them to a froth; add 1 cup sugar, 1 quart milk and set the saucepan in a vessel of boiling water over the fire; stir constantly until the custard nearly boils; then remove it from the fire and set the saucepan in cold water; when cold strain it through a sieve, add 2 teaspoonfuls vanilla or lemon extract, put the custard in the freezer and freeze as directed.

268.**Plain Ice Cream.**— Put 2 teaspoonfuls cornstarch, 6 eggs, 2 cups sugar and 2 quarts milk in a saucepan over the fire and stir till just about to boil; remove from the fire, flavor with lemon or vanilla and finish as directed.

269.**Plain Ice Cream (another way).**— Put 1 quart milk and 1 quart rich, sweet cream with the yolks of 8 eggs and 2 cups sugar over the fire and stir till just about to boil; remove from fire, beat 71 the whites of the 8 eggs to a stiff froth and add them to the custard; add 3 teaspoonfuls vanilla and finish as directed.

270.**Pistachio Ice Cream.**— ¼ pound blanched pistachio nuts, ¼ pound blanched almonds, 1 quart rich, sweet cream, 1½ cups sugar, the yolks of 8 eggs and 1 teaspoonful vanilla extract; pound the nuts with a little water very fine; place a saucepan over the fire with the cream, the yolks of the 8 eggs, sugar and vanilla and stir until nearly boiling; remove from the fire, stir in the nuts and when cold press the whole through a sieve; finish as directed. Almond ice cream is made the same way.

271.**Maraschino Ice Cream.**— Place a saucepan with 1 quart cream, ¾ pound sugar and the yolks of 6 eggs over the fire and stir till it nearly boils; remove from fire, strain through a sieve and when cold add 1½ gills maraschino; finish as directed. Rum ice cream is made in the same manner.

272.**Caramel Ice Cream.**— 1½ cups sugar, the yolks of 7 eggs, 1 quart sweet cream and 1 tablespoonful orange blossom water; boil ½ cup sugar with ¼ cup water until it turns to a light brown color, add ¼ cup boiling water and stir till the sugar is dissolved; put it in a saucepan with the cream, 1 cup sugar, yolks and orange water and stir the whole over the fire until nearly boiling; when cold strain it through a sieve and finish as directed.

273.**Tea Ice Cream.**— 1 ounce of the very best tea, 1 quart cream, the yolks of 6 eggs and ¾ pound sugar; boil the cream, put in the tea, cover and let it stand 5 minutes; strain through a sieve and when nearly cold mix the cream, yolks and sugar together and stir over the fire until nearly boiling; remove from fire and when cold finish as directed.

274.**Coffee Ice Cream, No. 1.**— 1 quart cream, 1 pint milk, 2 cups sugar, the yolks of 6 eggs and 2 ounces freshly ground coffee; boil the milk, put in the coffee, cover and set it aside to cool; next put the cream, yolks and sugar in a saucepan and stir over the fire till it nearly boils; remove from fire, add the coffee and when cold strain through a fine sieve, finishing as directed.

72

275.**Coffee Ice Cream, No. 2.**— ¾ pound sugar, 1 quart sweet cream, the yolks of 6-8 eggs and 5 ounces unroasted Mocha coffee; roast the coffee in a pan over the fire and put it into half of the boiling cream; cover and let it stand till cold; put the remaining cream, yolks and sugar in a saucepan over the fire and stir till nearly boiling; remove from the fire, add the coffee cream with the beans and let it stand till cold; then strain through a sieve and freeze as directed. 2 ounces freshly ground coffee may be used instead of the beans.

276.**Ice Cream (Simple).**— 1 quart sweet cream, 1 cup sugar and 2 teaspoonfuls vanilla or lemon extract; mix this well together, pour into a freezer and finish as directed. Or take equal parts of cream and milk; to 1 quart of this add 1½ cups sugar and any flavoring that may be desired; pour into the freezer and finish as directed.

277.**Plain Chocolate Ice Cream.**— 2 pints cream, 1 pint milk, 2 cups sugar, ¼ pound Baker's grated chocolate and 2 teaspoonfuls vanilla extract; place a saucepan with the milk, chocolate and cream over the fire, add the sugar, stir and boil for a few minutes; remove from fire and when cold freeze as directed.

278.**Nut Ice Cream.**— ½ pound blanched walnuts, the yolks of 6 eggs, 1½ cups sugar, 1 quart cream and 1 teaspoonful vanilla; pound the walnuts fine; put the cream, yolks, sugar and vanilla in a saucepan and stir over the fire till nearly boiling; remove from the fire, add the nuts and when cold strain it through a sieve; freeze as directed.

279.**Fine Chocolate Ice Cream.**— ½ pound grated chocolate, ½ pound sugar, 1 quart sweet cream, the yolks of 8 eggs and 1 teaspoonful vanilla extract; place a saucepan with ½ pint cream and the chocolate over the fire and stir and boil till chocolate is dissolved; stir the yolks, sugar, the remaining cream and vanilla together, add it slowly to the chocolate and continue stirring until nearly boiling; remove from fire and finish as directed.

73

280.**Strawberry Ice Cream.**— 1 pint ripe strawberries, 1 pint rich, sweet cream, 1 pound sugar and 1 teaspoonful vanilla extract; wash and drain the berries, mash them fine and mix with the sugar; cover and let stand till sugar is melted; press them through a sieve, mix the strawberry pulp with the cream and vanilla and put the whole into a freezer and freeze as directed. Raspberry, peach and apricot ice creams are made the same way.

281.**How to Make Ice Cream Without a Freezer.**— The process is so easy of manipulation and the expense incident thereto so small that most anybody can prepare it without any great trouble. All that is necessary for its preparation is a butter tub or a large pail, some ice, rock salt, atin form with tube in the center and a cover that fits it closely. The ice is best broken to pieces by putting it into a coarse bag and pounding with a hatchet. By this process no ice is wasted and there is no muss.

282.**Vanilla Ice Cream, No. 1.**— Set a plain tin form with tube in the center into cracked ice and salt; place a saucepan with 1 quart milk, the yolks of 6 eggs, 1 cup sugar and 1 tablespoonful cornstarch over the fire and stir with an egg beater till nearly boiling; remove from the fire, set saucepan in cold water and continue stirring till cold; then add the whites of the 6 eggs, beaten to a stiff froth, and 2 teaspoonfuls vanilla extract; pour this into the form, put on the cover and paste a strip of buttered paper around its edge, to prevent the salt water from entering; put a thick layer of cracked ice in the bottom of a butter tub and sprinkle over it a handful of rock salt; set the form onto the ice and fill the space between it and tub with cracked ice and salt; lay a thick layer of ice on top of form and sprinkle with salt; cover the tub with a carpet or bag and let it stand in a cool place for 4 hours; when ready to serve take the form out of the ice, remove the paper, dip the form into hot water, quickly wipe dry, turn the cream onto a dish and serve.

283.**Vanilla Ice Cream. No. 2.** — Place a deep kettle into cracked ice and put into it 1 quart rich, sweet cream; beat this with an egg beater until thick and add 1 cup powdered sugar and 2 teaspoonfuls 74 vanilla extract; put the cream into a tin form with a tube in the center, cover tightly, paste a strip of buttered paper around the edge of cover and finish the same as in foregoing recipe. NOTE. — For chocolate cream dissolve ¼ pound grated chocolate in ½ cup water, let it boil for a few minutes and when cold stir into the whipped cream prepared as above. Preserved peaches cut into small pieces or preserved pineapples cut into dice and mixed with the whipped cream is very nice. 1 dozen macaroons pounded fine and mixed with the whipped cream is also excellent. Pumpernickel cut in slices, dried in an oven and rolled fine may also be used. Candied fruit cut into pieces and fresh or preserved strawberries, as also cherries, apricots and oranges, can be used in the same way. For a small family 1 pint of cream will be sufficient.

284.**Fruit Ice Cream.** — Stir 1 quart cream with the yolks of 6 eggs and 1½ cups sugar over the fire till it nearly boils; remove from fire and when cold put the cream into the freezer and work till half frozen; then add any kind of fruit — either fresh strawberries or preserved pineapple cut into dice, ripe peaches cut into quarters, preserved pitted cherries or apricots; then finish as directed. The fruit may also be stirred into Custard Ice Cream in the same manner.

285.**Fruit Ice.**— The principal point in making fruit ice is to use the exact quantity of sugar. If the mixture contains too much sugar it will not freeze; if too little sugar the ice will be hard and dry. The better way is to try a little of it before putting the whole mixture into a freezer. If hard and dry add some thick sugar syrup; if it does not freeze at all add some cold water or a very thin syrup of sugar.

286.**Cold Sugar Syrup for Fruit Ice.**— Dissolve 1 pound sugar in 1 pint cold water and use as directed in following recipe. This is the ordinary syrup of 32 degrees used for fruit ice. If a thicker syrup is wanted dissolve 1 pound sugar in ½ pint water.

287.**Strawberry Ice.**— Wash and drain 1 quart ripe strawberries and press them through a sieve; mix the pulp with 1 pint 75 sugar syrup, as in No. 286, and the juice of 2 lemons; press it through a fine hair sieve, put it into a freezer and freeze as directed.

288.**Pineapple Ice.**— Choose a large, ripe pineapple, pare and grate it, or cut into pieces, and chop fine; put the pulp into a porcelain dish and pour over it ½ pint sugar syrup; cover and let it stand 1 hour; then add another ½ pint sugar syrup and the juice of 1 lemon; press it through a sieve and put in a freezer to freeze.

289.**Tutti Frutti Ice.**— Pound ¾ pound blanched sweet almonds and 12 bitter ones with a little cold water very fine; pour over 1 pint water and let them stand for ½ hour; then press them through a hair sieve; mix this almond milk with 2 pints sugar syrup and 1 teaspoonful vanilla; put this into a freezer and freeze; when frozen take the paddle of the freezer out and put in different kinds of fruit cut into small dice — either fresh or preserved peaches, pineapples, plums, cherries or apricots.

290.**Peach Ice.**— Pare and cut 12 large, ripe peaches into pieces, press them through a sieve, mix with a little over 1 pint sugar syrup and freeze. Ices from egg plums and apricots are made in the same way.

291.**Melon Ice.**— Choose a nice, ripe musk melon, cut it in half, remove the seeds and green portion and press the soft part through a sieve; mix it with an equal quantity of sugar syrup, alittle vanilla extract, the juice of 1 lemon, alittle orange blossom water and freeze as directed.

292.**Orange Ice.** — Mix 1 quart sugar syrup with the juice of 10 oranges, put in the thin peel of 2 oranges and let it stand for 6 minutes; remove the peel, pour the syrup through a sieve and freeze as directed.

293.**Lemon Ice.** — Mix the juice of 5 lemons with 1 quart sugar syrup and freeze.

294.**Sorbet** is served in Europe at balls and suppers.

295.**Champagne Sorbet.** — Dissolve 1 pound sugar in 1½ pints cold water, add the juice of 2 lemons and 6 oranges and a 76 little of the peel of each and let it stand 10 minutes; remove the peel, add ½ bottle champagne, put it into a freezer and work for ¼ hour; 10 minutes before serving add ½ bottle champagne, work it for a few minutes longer and then serve in glasses. Sorbet should not freeze hard; it should be a creamy liquid and ice cold.

296.**Pineapple Sorbet.** — 1 quart pineapple syrup, the juice of 3 oranges and 1 lemon; mix all together, strain and put into a freezer, work it for ½ hour and add by degrees during the freezing process ½ bottle champagne; finish the same as Champagne Sorbet. Sorbets of oranges, strawberries, peaches, cherries and apricot syrup are made in a similar manner.

297.**Strawberry Sherbet.** — Press 1 quart ripe strawberries through a sieve, add ¾ pound sugar dissolved in 3 pints cold water, add the juice of 1 lemon and 2 teaspoonfuls orange flower water, cover and let stand for 2 hours; then strain through a fine sieve and set on ice for 1 or 2 hours; serve ice cold in small glasses.

298.**Orange Granite.** — Mix 1 pint orange juice with 3 pints sugar syrup, as in No. 286, the juice of 2 lemons and the peel of 1; let it stand a few minutes then strain through a sieve; pour the mixture into a freezer, cover and turn for 5 minutes; then take off the cover, cut the frozen part loose from the sides of freezer, turn for a few minutes longer and serve. Granite must not be frozen hard; it should have little lumps all through it. Granites of strawberries, pineapples, raspberries, currants, peaches, apricots or cherries are made in a similar way. In granite of currants omit the lemon juice.

299.**Spongada aux peches.** — Pare and cut into pieces some ripe peaches, press them through a sieve and take for 1 pint peach pulp

1 pint sugar syrup, as in No. 286, 2 teaspoonfuls vanilla extract, 1½ gills white of egg not beaten and 10 bitter almonds pounded to a paste with a little water; mix all well together and strain twice through a sieve; pour this into a freezer, cover and turn for 5 minutes; take off the cover, cut the frozen part loose from the sides of freezer, cover and turn again; repeat the operation of cutting 77 from the sides every 5 minutes; as soon as the mixture begins to thicken remove the paddle of freezer, work the mixture up and down with a large spoon and press it towards the sides and bottom of freezer; as soon as the contents of freezer have increased to double their size add 2 tablespoonfuls maraschino and serve in glasses.

300.**Spongada au chocolat.**— 1 pint sweet cream, ½ pound finely grated chocolate, ½ pound sugar, 2 teaspoonfuls vanilla extract, 1½ gills white of egg and 1 large cup water; boil the chocolate in the water for 5 minutes; when cold mix all the ingredients together and finish the same as in preceding recipe.

301.**Spongada au Café.**— 1 pint cold sweet cream, ½ pint very strong coffee, ½ pint whites of eggs (not whipped) and 1 pound powdered sugar; mix all together and finish the same as Spongada aux Pêches.

302.**Spongada au marasquin.**— 1½ gills white of eggs, 2 pints rich, sweet cream, 1½ cups sugar, 2 teaspoonfuls vanilla extract and 1½ gills maraschino; dissolve the sugar in the cream, add vanilla and the white of egg without having been beaten and finish the same as Spongada aux Pêches. The maraschino is to be added shortly before serving.

303.**Orgeat of Almond Milk.**— 1 pound sweet and 12 bitter almonds are scalded in boiling water, freed from their brown skins and laid for 1 hour in cold water; drain the almonds on a sieve and pound them fine with 1 pound sugar and a few spoonfuls water; put the pounded nuts into a porcelain dish, pour over them 4 quarts cold water, add 2 teaspoonfuls vanilla extract and strain through a napkin. The napkin should be well washed in cold water and wrung out previous to being used. Put this almond milk into glass bottles and place them on ice before serving.

304.**Thé Polonaise.**— Place a porcelain-lined saucepan over the fire with 1 bottle Rhine wine, 2 bottles weiss beer, the rind of 1 lemon, apiece of stick cinnamon, 2 whole eggs, the yolks of 6 78 and sugar to taste; beat the whole with an egg beater over the fire till nearly boiling; instantly remove, continue beating for a few minutes longer and serve hot in cups. This is served at the end of a ball or party shortly before the guests go home.

305.**Iced Tea.** — Boil 1 quart milk with 4 tablespoonfuls sugar, add 1½ ounces tea, cover and set aside for 5 minutes; then strain and when cold pour it into an ice form; finish with whipped cream the same as Coffee Ice.

306.**Iced Coffee.** — Boil 1 quart milk with 4 tablespoonfuls sugar, add 1 cup coarsely ground coffee, cover and let it stand for 15 minutes; then strain and when cold put it into an ice form, cover and set into cracked ice with a little rock salt sprinkled between; let it stand for ½ hour; then thoroughly stir it with a long-handled spoon and mix with 1 pint whipped cream; serve in small cups.

307.**Bread Crumbs.** — Take stale bread or pieces which are left from the table, put them in a long, shallow tin pan and place in a medium hot oven; leave the door of oven open a little, so that the bread may dry slowly; when it is dry and has become a delicate brown color put the bread on a pastry board and roll it fine with a rolling pin; sift the crumbs through a sieve, return those which remain in it back on the board and roll and sift again; continue in this way until all the crumbs have been rolled fine and sifted; put them into a jar or box until wanted.

308.**How to Prepare a Pudding Form.** — Rub the inside of a form well with butter and thickly sprinkle it with fine bread crumbs; turn the form upside down, in order that the loose crumbs may fall out; the cover of the form must be treated the same way; fill form with the pudding mixture, put on the cover and tie it firmly with a cord; set the form in a vessel of boiling water so that ⅓ of it is immersed; then cover the vessel and boil slowly till done; add more water according as it diminishes through boiling. The form may be put in a large saucepan of boiling water and the latter covered with a deep dish or pan; but care must be taken not to 79 have too much water in the saucepan, otherwise it will get inside of the form.

BOILED AND BAKED PUDDINGS.

Half the quantity of any of the following recipes will be sufficient for a small family, but care must be taken in measurement to use only the *exact* half.

309.**Plum Pudding.**— Take ¾ pound finely minced suet, ½ pound stoned raisins, ½ pound well cleansed currants, ¼ pound finely cut citron, 5 well beaten eggs, the grated rind of 1 lemon, 1 grated nutmeg, 2 teaspoonfuls ground cinnamon, 1 teaspoonful cloves, 1 teaspoonful salt, 2 cups bread crumbs, ½ cup sour cream or milk, 1 cup syrup, 1 cup brown sugar, 1¼ pounds sifted flour, 1 teaspoonful baking soda dissolved in a little boiling water, 2 teaspoonfuls cream of tartar mixed with the flour and 1 glass brandy; mix all well together; have ready a large pudding form, rub the inside well with butter and sprinkle with bread crumbs; fill the mixture into the form and boil 4 hours; when done turn the pudding out onto a dish, pour brandy or rum over it, light and bring the pudding to table while burning; serve with hard sauce made as follows:—Stir 2 tablespoonfuls butter with 8 tablespoonfuls powdered sugar to a cream, add the yolks of 2 eggs, 4 tablespoonfuls brandy, alittle nutmeg and the beaten whites of 2 eggs; sufficient for 20 persons. If any of the pudding be left put in a stone jar and it will keep for a long time. When wanted cut off a piece sufficient for dinner, put it in a colander over a vessel of boiling water, cover with a plate, steam for ½ hour and serve. The quantities cited in this recipe will make 1 large pudding or 2 medium sized ones.

310.**English Plum Pudding.**— 1½ pounds Muscatel raisins, 1¾ pounds currants, 1 pound Sultana raisins, 2 pounds sugar, 2 pounds bread crumbs, 16 eggs, 2 pounds finely chopped suet, 6 ounces finely cut citron, the grated rind of 2 lemons, 1 ounce 80 ground nutmeg, 1 ounce cinnamon, ½ ounce ground bitter almonds and ¼ pint brandy; stone and cut up the raisins, but do not chop them; wash and dry the currants; mix all the dry ingredients together and moisten with the eggs, which should be well beaten; stir in the brandy and when all is well mixed butter and flour a strong pudding cloth; put in the mixture, tie up cloth very tightly, put into a large vessel of boiling water and boil from 6 to 8 hours; serve with brandy sauce. This quantity may be divided and boiled in buttered moulds. For small families this is the most desirable way, as the above ingredients will be found sufficient to make a pudding for 25 persons. This pudding is excellent, but any one troubled with dyspepsia had better not eat it.

311.**Biscuit Pudding.** — 2 cups milk, 1 cup butter, 2 cups flour, 1 cup sugar, 10 eggs, 1 teaspoonful vanilla extract and the grated rind and juice of 1 lemon; put the milk with ½ of the butter over the fire; as soon as it boils stir in the sifted flour and keep on stirring until the contents of saucepan form into a smooth paste and loosen from bottom of saucepan; then transfer it to a dish and set aside to cool; stir the remaining butter to a cream and add alternately the yolks of eggs, the sugar and the paste; thoroughly stir this and add the lemon, vanilla and the 10 whites beaten to a stiff froth; fill into a well buttered and floured form, boil 2 hours and serve with wine cream sauce. NOTE. — This pudding should be served as soon as taken from the form. The above ingredients will make a pudding sufficient for 10 persons.

312.**Cottage Pudding (baked).** — Take 1½ cups milk, 3 cups prepared flour, ½ cup sugar, ½ cup butter, 3 eggs and the grated rind of 1 lemon; stir butter and sugar to a cream, add by degrees the eggs and lemon and lastly, alternately, the flour and milk; butter a long tin pan, sprinkle with bread crumbs, pour in the mixture and bake ½ hour; serve with wine or nutmeg sauce; in serving cut the pudding into squares; sufficient for 10 persons.

313.**Cottage Pudding (boiled).** — Prepare a batter the same as in foregoing recipe, butter a pudding form, sprinkle with bread crumbs and pour in the mixture; the form should be about ¾ full; boil 2 hours and serve with following sauce: — Stir ½ cup butter with 1½ cups powdered sugar to a cream, add the yolk of 1 egg, some pitted or preserved cherries and 1 tablespoonful brandy; or add bruised strawberries, blackberries or peaches cut into small pieces.

314.**Cottage Apple Pudding.** — Prepare a batter as for Cottage Pudding (baked) and add 3 cups finely cut apples; in other respects treat the same as foregoing recipe and serve with lemon sauce.

315.**Batter Fruit Pudding.** — ¼ pound butter, 4 tablespoonfuls sugar, 4 eggs, 2 cups milk, 4 cups prepared flour, 1 cup seedless raisins and currants, ½ cup finely cut citron, the grated rind of 1 lemon and a little nutmeg; stir butter and sugar to a cream and add the eggs by degrees; then add alternately the sifted flour and milk, next the fruit, lemon and nutmeg; butter a pudding form, sprinkle with bread crumbs, put in the mixture and boil 2 hours; serve with hard, brandy or punch sauce. NOTE.—The fruit should be dusted with flour before adding it to the batter; sufficient for 10 persons.

316.**Prince Regent Pudding.**— After removing the crust off a 5 cent loaf of stale bread grate on a grater and pour 1 pint milk over it; then stir 2 tablespoonfuls butter with 4 tablespoonfuls sugar to a cream, add the yolks of 7 eggs, 4 tablespoonfuls finely cut citron, ¼ pound well cleansed seedless raisins, the bread and the beaten whites of the eggs; fill this into a pudding form which has been well buttered and sprinkled with bread crumbs, close tightly and boil 2 hours: serve with sherry wine, cream or brandy sauce.

317.**Layer Pudding (German style).**— Cut a 5 cent Vienna loaf of bread (after the crust has been removed) into thin slices; butter these on both sides, dip each slice into milk, lay them on top of one another and set aside; mix together ½ cup stoned raisins, 3 tablespoonfuls well cleansed currants, 2 tablespoonfuls finely cut citron and ½ teaspoonful cinnamon; beat 8 eggs to a froth and add, stirring constantly, 1 pint milk; next butter a 82 pudding form and sprinkle thickly with bread crumbs; put in a layer of the slices of bread, sprinkle over them some of the fruit mixture and 2 tablespoonfuls sugar; then put in another layer of bread, fruit and sugar; continue until all is used; then pour over it the milk and eggs, cover the form closely and boil 1½ hours; serve with hard or cherry sauce; sufficient for 12 persons.

318.**Portugal Pudding.**— Grate the crust from a small loaf of bread and soak the latter in milk; when soft press it out and put in a saucepan with 1 tablespoonful butter and 1 tablespoonful clarified dripping; stir for 5 minutes over the fire, transfer it to a dish and as soon as cold mix with the yolks of 6 eggs, 5 tablespoonfuls sugar, ¼ pound stoned raisins, ¼ pound well washed currants, the grated rind and juice of 1 lemon, 2 tablespoonfuls finely cut citron, ½ cup Cognac or rum and lastly the beaten whites of the eggs; butter a pudding form, sprinkle with fine bread crumbs, fill in the mixture, close tightly and boil 2 hours; serve with hard or wine sauce; sufficient for 10 persons.

319.**Ipsilanti Pudding.**— Mix 1 cup bread crumbs with 1 cup sweet cream and let it stand ½ hour; stir ¼ pound butter with 6 tablespoonfuls sugar to a cream and add by degrees the yolks of 8 eggs; after this is well blended together add by degrees the bread, the grated rind of 1 lemon, 6 ounces finely cut citron, 4 tablespoonfuls bread crumbs fried in butter, 1 teaspoonful ground cinnamon and if handy 2 tablespoonfuls finely cut preserved ginger; beat the whites of the 8 eggs to a stiff froth, mix all well together, fill the mixture into a pudding form which has been well buttered and sprinkled with bread crumbs, boil 2 hours and serve with wine or cherry sauce.

320.**Fine Cherry Pudding (of fresh fruit, for a family of 6).**— ¼ pound finely chopped suet, ½ pound flour, 2 eggs, 2 tablespoonfuls sugar, ½ pint milk or water, 1½ teaspoonfuls baking powder sifted with the flour and a little salt; mix all the ingredients together; add ½ pound cherries (minus the pits) to the batter and fill the mixture into a pudding form which has been well 83 buttered and sprinkled with bread crumbs; boil the pudding 2 hours and serve with the following sauce:—Stir 2 tablespoonfuls butter with 8 of powdered sugar to a cream, add the yolks of 2 eggs, 2 tablespoonfuls Cognac, rum or sherry wine and lastly the whites of the eggs beaten to a stiff froth and ½ cup stoned cherries. 2 tablespoonfuls of lard, butter or clarified drippings may be substituted for suet, and instead of cherries any other kind of fruit may be used.

321.**Cherry Pudding (of preserved Cherries).**— ¼ pound finely chopped suet, 2 cups flour, 1½ teaspoonfuls baking powder, 2 tablespoonfuls sugar, ½ teaspoonful salt, 2 eggs and 1 cup milk or water; sift flour, sugar, salt and powder into a bowl and mix them with the finely chopped suet; make a hole in center, put in the yolks of the 2 eggs, gradually add the milk and mix the whole into a smooth batter; lastly add the whites of the eggs, beaten to a stiff froth; put a can of preserved cherries in a colander or sieve, drain off all the liquor and stir 1 cup of the cherries into the batter; butter well with butter or lard a pudding form and dust it with finely sifted bread crumbs; fill in the mixture, put on the cover and tie it tightly with a string;

place the form in a large saucepan of boiling water (the form should not be immersed in the water more than half its depth), cover the saucepan with a deep dish or pan, so that no steam can escape, and boil 2 hours; according as the water boils away add more boiling water; when done turn the pudding onto a round plate and serve with the following sauce:—Put 1 tablespoonful cornstarch in a saucepan and mix it with a little cold water; add 1 cup boiling water, stirring constantly, and let it boil for 2 minutes; then remove it from fire, add 1 cup cherry syrup, 1 teaspoonful vanilla, alittle more sugar, 1 glass sherry wine and lastly 2 tablespoonfuls preserved cherries. 1 tablespoonful lard or butter may be used instead of suet. This pudding can be made of all kinds of preserved fruit; sufficient for a family of 6 persons.

322.**Cherry Batter Pudding.**— Stir ½ cup butter with 3 table-spoonfuls sugar to a cream, add by degrees 4 eggs and alternately 4 cups Hecker's prepared flour and 2 cups milk; then add 1 quart cherries; butter a pudding form, sprinkle with bread 84 crumbs, put in the mixture, cover the form, set in a kettle of boiling water, so that the form is half immersed, and boil 2 hours; serve with cherry, hard or wine sauce; or stir one cup pitted cherries into the hard sauce. NOTE.—If fresh cherries are not available the California canned cherries may be used, and will be found excellent. If canned fruit is used drain off the juice and only put the cherries into the batter, using the liquor either for the sauce or to make a form of jelly (see Jelly). California preserved peaches and apricots also make very fine puddings. The above recipe is sufficient for 12 persons.

323.**Plain Suet Pudding.**— ½ pound finely chopped suet, 4 cups sifted flour mixed with 3 teaspoonfuls baking powder, 1 teaspoon-ful salt, ½ cup sugar, 4 eggs and 1 pint milk; beat up the eggs and add the salt and milk; when this is well beaten together add the flour with the powder and sugar, mix the suet with a little flour and stir it into the batter; butter a pudding dish, sprinkle with bread crumbs, pour in the mixture, put on the cover, set the form in a kettle of boiling water, so that the water covers half of the form, and boil 2 hours; serve with strawberry sauce made as follows:—Stir 2 tablespoonfuls butter with 1 cup powdered sugar to a cream; wash and mash finely 1 cup strawberries and mix them with the sauce; stir a handful of whole berries into it, put the sauce into a glass dish, smooth it with a knife and set some whole strawberries all around the top. Pitless cherries, cut up peaches, pitless plums or blackberries may be substituted for strawberries. The above quantities will make a pudding sufficient for 12 persons.

324.—**Suet Pudding (with Apples).**— Dust 3 cups finely chopped apples with flour and stir them into the plain suet pudding mixture; otherwise treat the same as Plain Suet Pudding and serve with hard or sherry wine sauce.

325.**Blackberry Pudding** is made in the same manner as Plain Suet Pudding, except that 1 quart well washed and floured blackberries are stirred into the batter; serve with hard sauce, into which 1 cup bruised blackberries may be stirred. Huckleberry pudding is made the same way.

85

326.**Cherry Suet Pudding.**— Add to the plain suet pudding mixture 1 pound stoned cherries (which should be dusted with flour before adding) and finish the same as Apple Suet Pudding; serve with following sauce:—Take 1 pound cherries and pound half of them fine in a mortar; place the whole cherries with the pounded ones in a saucepan over the fire, add 1 cup water and boil till tender; then strain them through a sieve, return the liquor to saucepan, sweeten to taste, add 2 teaspoonfuls cornstarch dissolved in a little cold water, apiece of cinnamon and boil a few minutes; then add ½ pint claret and serve; or stir into the hard sauce 1 cup pitted cherries. Both of these sauces are excellent with cherry pudding.

327.**Suet Pudding (with Nuts).**— Stir into the plain suet pudding mixture 1 cup chopped almonds, walnuts or any kind of nuts; boil in the form the same as Plain Suet Pudding and serve with nut sauce, which is made as follows:—Stir ½ cup butter with 1½ cups powdered sugar to a cream and add the yolks of 2 eggs and 1 cup chopped nuts.

328.**Suet Pudding (with Raisins).**— Stir into the plain suet mixture 1½ cups stoned raisins broken into pieces, boil the same as Plain Suet Pudding and serve with hard sauce flavored with rum and mixed with ½ cup blanched almonds or walnuts broken into pieces.

329.Suet Fruit Pudding.— 1 cup finely chopped suet, 1 cup milk, 1 cup molasses, 1 cup seedless raisins, 1 cup currants, 1 teaspoonful cinnamon, ½ teaspoonful cloves, ½ nutmeg, ½ teaspoonful salt, 4 eggs, 2 cups bread crumbs and 2 cups sifted prepared flour; mix all the ingredients together, fill the mixture, into a well buttered pudding form, boil 2½ hours and serve with the following sauce:—Boil 1½ cups water, add 1 tablespoonful flour wet with ½ cup cold water and boil for a few minutes; then add 1 tablespoonful butter, alittle nutmeg, the juice of 1 lemon and sweeten to taste.

330.Marrow Pudding.— ¼ pound finely chopped beef marrow, ¼ pound finely chopped suet, 5 eggs, 2 cups bread crumbs, ½ 86 cup milk, ½ pound prepared flour, ½ cup rum, ½ cup sugar, ¼ pound raisins, the same of well washed currants, 2 ounces finely cut citron, the grated rind of 1 lemon, ½ grated nutmeg and 1 teaspoonful salt; mix all together with the yolks of 5 eggs and add lastly the whites of the eggs, beaten to a stiff froth; put the mixture in a well buttered pudding form and boil 3 hours; serve with hard or brandy sauce. This pudding may also be boiled in a cloth, but is much finer when done in a form; sufficient for 10 persons.

331.Fig Pudding.— ½ pound finely chopped suet, 4 eggs, 1 pint milk, ½ pound figs cut into pieces, 1 pound flour and 3 teaspoonfuls baking powder; mix flour and baking powder together, add suet, eggs, 1 teaspoonful salt, the figs and mix it with the milk into a stiff batter; add 2 tablespoonfuls sugar, fill the mixture into a well buttered pudding form and boil 2 hours; serve with hard or wine sauce.

332.Apple Pudding (baked).— Stir 2 tablespoonfuls butter to a cream, add ¼ pound sugar, ½ cup chopped almonds, the yolks of 6 eggs, 3 tablespoonfuls flour, the grated rind of 1 lemon and 1 quart stewed apples; mix all together, add the beaten whites of the eggs, fill the mixture into a buttered pudding dish and bake 1 hour; when done sprinkle with powdered sugar and serve without sauce in the same dish in which it was baked.

333.Pineapple Pudding (or Souflée).— Boil 1 pint milk with 1 tablespoonful butter, while boiling sprinkle in 1 pint sifted flour and stir constantly until it has formed into a smooth dough and loosens itself from bottom of saucepan; transfer it to a dish to cool; stir 1 tablespoonful butter to a cream, add alternately the yolks of 4 eggs, 4 tablespoonfuls sugar, 1 cup milk, the grated rind of 1 lemon and the paste (1 spoonful at a time); lastly add the whites of the eggs, beaten to a stiff froth; butter a deep pudding dish and sprinkle with bread crumbs; put in a layer of the mixture and sprinkle over it a few bread crumbs; put over this a layer of stewed or preserved pineapples (cut into small dice) and sprinkle over a few bread crumbs; then a layer of the mixture and pineapples, 87 until all is used; let the last layer be the mixture; bake 1 hour and serve with raspberry sauce; sufficient for a family of 8. Preserved peaches, apricots or cherries may be used instead of pineapples.

334. Almond Sponge Pudding. — Place a saucepan with 1 pint milk and 1 tablespoonful butter over the fire; as soon as it boils stir in ½ pound sifted flour, keep stirring until it forms into a smooth dough and loosens itself from bottom of saucepan; then transfer it to a dish and set aside to cool; stir 1 tablespoonful butter to a cream and add alternately the yolks of 9 eggs, 9 tablespoonfuls sugar and the dough; stir it with a potato masher until all the dough, the 9 yolks and 9 tablespoonfuls sugar have been used; add 1 cup finely chopped or grated almonds, the juice and rind of 1 lemon and lastly the beaten whites of 9 eggs; fill this mixture into a pudding form which has been well buttered and sprinkled with bread crumbs or flour and boil 2 hours; serve with the following sauce: — Place a saucepan over the fire with 1 pint white wine, 4 tablespoonfuls sugar, alittle lemon rind and 3 whole eggs; beat this until just about to boil; instantly remove from the fire and serve in a sauciere with the pudding. NOTE. — The pudding should be served immediately after being turned out.

335. Nut Pudding. — Remove the shells from 1 pound walnuts, scald the nuts in boiling water and remove the fine brown skin; pound them in a mortar with white of egg and mix them with ¾ cup milk; boil ½ pint milk with ½ tablespoonful butter and while boiling add slowly 1 cup sifted flour; stir until it forms into a smooth paste and loosens itself from bottom of saucepan; put the paste in a dish, mix it with the pounded nuts and set aside to cool; stir 1½ tablespoonfuls butter to a cream and add by degrees the yolks of 8 eggs, 4 tablespoonfuls sugar and (by spoonfuls) the paste; when all is well mixed add the whites of the eggs, beaten to a stiff froth; butter a pudding dish, sprinkle with bread crumbs, fill in the mixture, set the dish into a pan of hot water and bake 1 hour in a medium hot oven; when done turn it onto a dish and serve with fruit or nut sauce; care should be taken not to use too small a dish, as the pudding raises very light; serve as soon as baked.

88

336.**Uncle Tom's Pudding.**— Mix 2 teaspoonfuls baking powder with 3 cups sifted flour and a little salt, add 1 cup molasses, 1 cup finely chopped suet, 2 tablespoonfuls sugar, 1 teaspoonful ground ginger, 1 teaspoonful cinnamon, ½ teaspoonful cloves, ½ grated nutmeg, 1 cup buttermilk and 3 eggs; butter a pudding form, sprinkle with bread crumbs, fill in the mixture and boil 2 hours; serve with lemon or hard sauce.

337.**Plain German Flour Pudding.**— Sift 4 cups flour, add ½ teaspoonful salt, 4 tablespoonfuls sugar, the grated rind of 1 lemon, 1 cup seedless raisins, 1 yeast cake dissolved in ½ cup warm milk, 2 tablespoonfuls melted butter, 1½ cups warm milk and 2 eggs; mix all together into a stiff batter; butter a pudding form, sprinkle with bread crumbs, fill in the mixture and set in a warm place till it rises to double its height; then cover the form and boil 2 hours; serve with roast meat and stewed fruit or with sauce.

338.**The Queen of Puddings (with Strawberries).**— 1 cup sugar, 2 cups fine bread crumbs, 6 eggs, 2 tablespoonfuls butter, the grated rind of 1 lemon, 4 cups milk and 1 pint strawberries; soak the bread crumbs in the milk for ½ hour; stir butter and sugar to a cream, add by degrees the yolks of the eggs and next the bread crumbs (by spoonfuls), stirring constantly; lastly add the whites of 3 eggs, beaten to a stiff froth, and the lemon; fill this into a buttered pudding dish, which should be a large one and but ⅔ full; bake until done; draw to the front of oven, put a layer of fresh strawberries over it, sprinkle with sugar and cover with a meringue made of the 3 remaining whites of eggs and 1 tablespoonful powdered sugar; put it back in the oven and bake for a few minutes, until the meringue begins to color; serve cold with cream or vanilla sauce. Any kind of fruit may be used instead of strawberries, as may also jelly or marmalade.

339.**Indian Pudding (boiled).** — Bring 1 pint milk to a boil, stir into it 1 cup yellow Indian meal and boil 5 minutes, stirring constantly; then take it from the fire and mix with 1 cup molasses, 1 tablespoonful ground ginger, 1 cup chopped suet, ½ teaspoonful 89 salt and 2 eggs; when this is well blended together fill it into a buttered pudding form and boil 3 hours; serve with the following sauce: — Mix 2 teaspoonfuls cornstarch with a little cold water, add 1½ cups boiling water and boil a few minutes; then add ¼ teaspoonful salt, 1 tablespoonful butter, the juice of 1 lemon, 1 cup sugar, 1 teaspoonful vanilla and some grated nutmeg.

340.**Economical Boiled Pudding.** — 1 cup milk, 1 cup stoned raisins, 1 cup fine chopped suet, ½ cup molasses, ½ cup brown sugar, 3 cups flour, 1½ teaspoonfuls baking powder, 1 teaspoonful grated nutmeg, the same of cinnamon, ½ teaspoonful cloves. Mix all together and boil in a form 2 hours; serve with lemon or vanilla sauce.

341.**Graham Flour Pudding (also called Imitation Plum Pudding).** — Two large slices of bread, ½ cup milk, ¼ pound butter, 1 cup stoned raisins, 1 cup currants, ½ cup finely sliced citron, 1 cup molasses, 1 cup graham flour, 1 cup sugar, 1 glass brandy and 1 teaspoonful baking soda dissolved in a little hot water; mixed with the molasses, 3 eggs, 1 teaspoonful allspice, ½ grated nutmeg and ½ teaspoonful salt; break the bread into small pieces and put it with the milk into a bowl; stir butter and sugar to a cream; add the eggs, one at a time, stirring a few minutes between each addition; next add the spice; then, alternately, the bread, molasses and flour; when this is well mixed dust the fruit with flour and stir it into the mixture; butter a pudding form and dust with fine bread crumbs; put in the mixture, close the form and set it in a kettle of boiling water (only enough water to half cover the form should be used); cover the kettle and boil 3 hours; serve with brandy or hard sauce. Half of the above quantities will make a pudding sufficient for a family of 6 persons.

342.**Madeira Pudding.** — Pare the crust off a 6 cent loaf of bread; cut the bread into slices and dip each slice in Madeira wine; mix 5 tablespoonfuls sugar with ¼ pound finely cut preserved orange peel, alittle nutmeg and cinnamon; have ready a well buttered pudding form, which sprinkle with fine bread 90 crumbs; first put in a layer of bread and sprinkle over it some of the mixed sugar; then a layer of currant jelly; continue in this fashion until all is used up; lay 1 tablespoonful butter in small pieces on top; beat up 6 eggs with 1 pint cream or milk and pour it into the form over the bread; close the form and boil 1½ hours; serve with the following sauce:—Put 1 pint Madeira wine in a saucepan with 3 or 4 eggs, the peel of 1 lemon, apiece of cinnamon and ½ cup sugar; place over the fire and stir with an egg beater until nearly boiling; instantly remove and serve with the pudding. If the sauce is allowed to boil it will be spoiled.

343.**Almond Pudding.**— Stir 2 tablespoonfuls butter to a cream, add 5 tablespoonfuls sugar, the yolks of 6 eggs, 1 cup chopped almonds, the grated rind and juice of 1 lemon, 2 cups fine bread crumbs and lastly the whites of the eggs, beaten to a stiff froth; butter a pudding form, sprinkle with bread crumbs, cover tightly and boil 1½ hours; serve with wine or cream sauce.

344.**Boiled Bread Pudding.** — Stir 1 tablespoonful butter to a cream with 2 tablespoonfuls sugar and add by degrees the yolks of 4 eggs, the grated rind of ½ lemon, 2 cups bread crumbs, ½ cup milk, 2 ounces seedless raisins, the same quantity of well cleansed currants and 2 tablespoonfuls finely chopped almonds; add lastly the beaten whites of the eggs; butter a form, sprinkle with bread crumbs, fill it with the mixture, put on the cover and boil 1½ hours; serve with sherry wine or cream sauce.

345.**Zwieback Pudding, No. 1.**— Butter a form and sprinkle with bread crumbs; take ½ pound round zwieback, ¼ pound seedless raisins, the same quantity of well cleansed currants and chopped almonds; put a layer of zwieback into the form and sprinkle some of the fruit over it; continue in this way until all is used; then beat up 6 eggs with 6 tablespoonfuls sugar and add 2½ cups milk; pour this over the zwieback in the form, cover tightly and let it stand 1 hour; then boil 2 hours; serve with fruit, wine or hard sauce; sufficient for 10 persons.

91

346.**Cabinet Pudding.**— Stir ½ cup butter with 4 tablespoonfuls sugar to a cream and add by degrees the yolks of 8 eggs and the grated rind and juice of 1 lemon; cut the crust off a 5 cent loaf of bread, grate the white part and add it to the above mixture with 1½

cups milk, ¾ cup finely cut citron and the whites of the eggs beaten
to a stiff froth; in the meantime pour over ¼ pound vanilla wafers
and ½ pound macaroons, some Madeira or sherry wine and sprinkle with finely sifted bread crumbs; put a layer of the bread mixture, an inch in thickness, into the form and cover it with a layer of
macaroons and wafers; then bread again; continue in this way until
all is used, the last layer being the bread mixture; close the form
tightly and boil 2 hours; serve with wine cream or hard sauce; sufficient for 12 persons.

347.**Lemon Pudding (baked).**— Stir 1 cup butter to a cream and
add by degrees the yolks of 10 eggs, 2 whole eggs, the grated rind
and juice of 3 lemons, 1 cup finely chopped almonds, 1 cup sugar
and lastly the whites of the eggs, beaten to a stiff froth; line a pudding dish with rich pie crust, put in the mixture and bake 1 hour. Or
take ¼ pound stale sponge cake, broken into small pieces, the juice
of 4 lemons and the grated rind of 2, 1½ cups sugar, 1 pint cream,
alittle salt and nutmeg, the yolks of 6 eggs and the beaten whites of
3; put this into a pudding dish lined with pie crust and bake ½ hour.

348.**Zwieback Pudding, No. 2.**— Soak ½ pound zwieback in 1
pint milk; stir ¼ pound butter with 3 tablespoonfuls sugar to a
cream; add by degrees the yolks of 6 eggs, ¼ teaspoonful cinnamon
and 1 cup finely chopped almonds; add lastly the zwieback and the
whites of the eggs beaten to a stiff froth; put the mixture into a well
buttered pudding form and boil 1 hour; serve with wine sauce.

349.**Pumpernickel Pudding.**— Cut some stale pumpernickel into
slices and dry them in the oven; then lay on a board, roll fine and
sift them; take 1 cup pumpernickel crumbs, 4 eggs, 2 tablespoonfuls
dripping or ½ cup finely chopped suet, 5 tablespoonfuls 92 sugar, 1
teaspoonful cloves, the same quantity of cinnamon and the grated
rind and juice of 1 lemon; stir the yolks of eggs and sugar to a
cream; add by degrees the dripping, bread crumbs and other ingredients; add lastly the beaten whites of the eggs; fill this into a well
buttered form and boil 1½ hours; serve with lemon or brandy sauce;
sufficient for a family of 6 persons. This pudding is the equal of a
fine plum pudding.

350.**Vienna Pudding.**— Stir ¼ pound butter with 1 cup sugar to a
cream and add by degrees the yolks of 7 eggs, 2 whole eggs, the

grated rind of ½ lemon and the juice of 2; set this in a vessel of boiling water and stir over the fire till it begins to thicken; then remove it, stir until cold and add the whites of the eggs, beaten to a stiff froth; butter a pudding form and sprinkle it with fine zwieback crumbs; fill in the mixture, put on the cover, set the form in a kettle of boiling water, cover closely and boil slowly for 1 hour; in serving turn the pudding onto a warm dish and send wine cream or fruit sauce to table with it. This pudding should be served immediately upon being turned out of the form.

351.**Chocolate Pudding.**— Stir 2 ounces butter with 1 cup powdered sugar to a cream, add by degrees the yolks of 9 eggs and stir for 20 minutes; then add 2 ounces finely chopped almonds, the grated rind and juice of 1 lemon, ¼ pound grated chocolate and 6 ounces rye bread which has been dried in the oven and rolled fine with a rolling pin; add lastly a glass of Madeira wine or rum and the whites of the 9 eggs, beaten to a stiff froth; put the mixture into a well buttered form, boil 2 hours and serve with wine or punch sauce.

352.**Apple Pudding (with Almonds).**— Place a saucepan over the fire with 1 tablespoonful butter; add 1 soup plate finely cut apples, 2 tablespoonfuls well cleansed currants, the same quantity of seedless raisins and finely cut citron, ¼ pound finely chopped almonds, the grated rind of ½ lemon or orange and ¾ cup sugar; stir this over the fire until the apples begin to get soft, add ½ cup raspberry or currant jelly and set aside to cool; beat up the yolks 93 of 7 eggs, add by degrees ¼ pound finely rolled zwieback, the apples and lastly the whites of the eggs, beaten to a stiff froth; fill this into a well buttered pudding dish and bake ¾ hour in a medium hot oven; when done turn the pudding onto a dish, dust with sugar and serve without sauce; sufficient for 10 persons. It may also be served in the dish in which it is baked.

353.**Nudel Pudding.**— Prepare the nudels from the yolks of 2 eggs and sufficient flour to make a stiff dough; roll it out thin and cut into long strips about 1½ inches wide; lay 4 strips on top of one another and cut them as fine as possible; then drop them into boiling milk and boil 10 minutes; drain on a sieve, return the nudels to the fire, add ½ tablespoonful butter, 3 macaroons pounded fine, 1 tablespoonful currant or apple jelly and a glass of sherry wine; shake this several times over the fire, spread the mixture on buttered tins ¾ inch in thickness and set in a cool place; put 1 ounce finely chopped or pounded almonds in ½ pint milk, let it stand ½ hour, add 1 whole egg, the yolks of 6 and 3 tablespoonfuls sugar; when well mixed strain through a sieve and cut the nudels with a cake cutter into rounds; put them in rows over one another into a form which has been well buttered and sprinkled with bread crumbs, sprinkle some pounded macaroons between, pour the cream over it and place the form in a vessel of hot water; set it on the stove to simmer gently for 1 hour; when done carefully turn the pudding out onto a dish and serve with almond, cream or fruit sauce. These quantities are sufficient for a family of 8 persons.

354.Potato Pudding.— Boil 8 large potatoes with their skins in water until done; take from the water and set them for a few minutes in the oven to dry; then set them in a cool place; when cold remove the skins and grate the potatoes on a grater; use only that portion which falls behind the grater; this should make 1 quart of grated potatoes; stir ¼ pound butter with 1 cup sugar to a cream and add by degrees the yolks of 8 eggs, the grated rind of 1 lemon, ¼ pound blanched almonds well pounded and 2 tablespoonfuls dry farina; when this is well mixed add the potatoes and lastly the whites of the eggs, beaten to a stiff froth; butter a pudding dish, sprinkle well with bread crumbs, put in the mixture and cover tightly; set the form into a vessel of boiling water (use only enough water to half cover the form), cover the vessel closely and boil slowly for 2 hours; when done take the form from the water and set it for a few minutes in the oven; then carefully turn the pudding onto a round plate and serve with the following sauce:—Stir 2 tablespoonfuls butter with 1 cup powdered sugar to a cream and add the yolks of 2 eggs and 1 cup fresh strawberries (either stir them into the sauce whole or mash them). Fresh cherries freed of their pits or preserved cherries may be used in place of strawberries. The pudding may also be served with either wine, lemon or fruit sauce; it should be served as soon as taken from the form.

BREAD AND APPLE PUDDINGS.

355.**Plain Bread Pudding, No. 1.**— Lay 3 slices of a 5-cent loaf of bread (minus the crust) in a pudding dish and pour over them 1 quart cold milk; set the dish on the side of stove to heat gradually; when hot stir 2 eggs with 2½ tablespoonfuls sugar to a cream and add a little cold milk or water and 1 teaspoonful essence of lemon; stir this into the bread and milk; put ½ tablespoonful butter in small bits on top, grate over some nutmeg, bake in oven from 20 to 30 minutes and serve hot or cold without sauce.

356.**Plain Bread Pudding (baked), No. 2.**— Put 3 slices of bread (minus the crust) into a pudding dish and pour over them 1 quart boiling milk; cover the dish and let it stand until cold; then beat up the bread with a fork; stir 3 eggs with 3 or 4 tablespoonfuls sugar to a cream and mix it with the bread; flavor with lemon; put a few small pieces of butter with a little grated nutmeg on top and bake in the oven till thick; serve with lemon or nutmeg sauce. ¼ pound raisins or currants may be added if liked.

357.**Bird's Nest Pudding.**— Peel 6 good sized greening apples, remove the cores with an apple corer without breaking the 95 fruit, put them in a long, shallow tin pan, pour over 2 cups boiling water, cover with a pan of same size and let them boil on top of stove for 5 minutes; then drain off all the water and put 1 teaspoonful apple or currant jelly into each apple. For batter take 1 cup flour, 1½ tea-spoonfuls baking powder, 1 cup milk, 2 eggs, ¼ teaspoonful salt, 1 tablespoonful lard, butter or clarified dripping and 2 teaspoonfuls sugar; sift flour, powder, sugar and salt into a bowl; put in the butter or lard and chop it fine in the flour; make a hollow in center and put in the yolks of the 2 eggs; then add the milk gradually and mix the whole into a smooth batter; add lastly the whites of the eggs beaten to a stiff froth; pour the batter over the apples and bake ½ hour in a medium hot oven; serve as soon as done and send the following sauce to table with it:—Stir 1 tablespoonful butter with 4 tablespoonfuls powdered sugar to a white cream and add the yolk of 1 egg, 1 tablespoonful rum or brandy and a little nutmeg; lastly stir in the white of the egg, beaten to a stiff froth; in serving give to each individual an apple on a small plate and a large spoonful of sauce on each apple; sufficient for a family of 6. This pudding has the advantages of being healthy and excellent, while not being expensive.

358.**Bread Pudding (boiled).** — Soak ¾ pound stale bread (minus the crust) in water; when soft press it out either in a napkin or with the hands; melt 2 tablespoonfuls butter or clarified dripping in a saucepan, add the bread and stir over the fire till it has formed into a compact mass and loosens itself from bottom of saucepan; transfer the bread to a dish; stir the yolks of 6 eggs with 4 tablespoonfuls sugar to a cream and add them by degrees to the bread; add 2 cups well cleansed currants, ½ cup finely chopped almonds, the grated rind and juice of ½ lemon, alittle nutmeg and ½ teaspoonful cinnamon; add lastly the whites of 6 eggs, beaten to a stiff froth; butter a pudding form and sprinkle with bread crumbs; fill in the mixture, put on the cover, place the form in a vessel of boiling water, cover closely and boil 2 hours; serve with lemon, fruit or hard sauce.

96

359.**Bread Fruit Pudding.** — Pare off the crust from a 5-cent loaf of bread and cut the loaf into thin slices; spread the slices on both sides with any kind of fruit marmalade; butter a pudding form, sprinkle with bread crumbs and lay in the bread; stir 8 eggs with ½ cup sugar and the grated rind of 1 lemon until they foam; add by degrees 1 pint hot milk, stirring constantly; pour this over the bread, cover the form tightly and boil 1½ hours; serve with wine cream sauce.

360.**Bread Sponge Pudding.** — Boil 1 cup milk with 1 tablespoonful butter; stir in while boiling 1 cup sifted flour and keep stirring until it has formed into a smooth paste and loosens itself from bottom of saucepan; transfer the paste to a dish; stir 1 tablespoonful butter to a cream and add, alternately, the yolks of 5 eggs, 4 tablespoonfuls sugar, the paste, ¼ pound well cleansed currants, alittle nutmeg and grated lemon peel; pour ½ cup rum over 1 cup fine bread crumbs and add them to the above mixture with the beaten whites of the 5 eggs; butter a pudding form, sprinkle with bread crumbs, fill in the mixture and boil in a tightly covered vessel of water for 2 hours; serve with wine cream, fruit or hard sauce; sufficient for 8 persons.

361.**Hanoverian Pudding.** — Pare and quarter 6 large pippin or greening apples and cut them into fine slices; put them in a saucepan with 1 tablespoonful butter, ½ cup sugar, 1 tablespoonful well cleansed currants, 1 tablespoonful seedless raisins, 2 tablespoonfuls finely cut citron and the grated rind and juice of ½ lemon; stir this over the fire till the apples begin to get soft; add ½ cup raspberry or currant jelly and set aside to cool; soak ¼ pound bread (minus the crust) in water and press it out in a napkin; then place it in a saucepan with 1 tablespoonful butter and stir over the fire until the bread loosens itself from bottom of saucepan; stir 1 tablespoonful butter to a cream and add, alternately, the yolks of 5 eggs, 2 tablespoonfuls sugar and the bread (by spoonfuls); add lastly the beaten whites of the eggs; next butter a pudding dish and sprinkle with bread crumbs; put in a layer of bread mixture and over it a 97 layer of apples; continue in this way until all is used; bake 40 minutes; serve with or without sauce.

362.**Huckleberry Pudding (German style).**— Soak a 5-cent loaf of bread (minus the crust) in milk till soft; press it out, put in a saucepan with 1 tablespoonful butter and stir over the fire to a smooth paste; transfer it to a dish and set aside to cool; stir 1 tablespoonful butter to a cream and add (alternately) the yolks of 8 eggs, 5 tablespoonfuls sugar and the bread (by spoonfuls); when this is well mixed add 1 pint huckleberries and lastly the whites of the eggs, beaten to a stiff froth; fill this into a well buttered and floured pudding form, cover closely and boil in a kettle of water 2 hours; serve with hard or wine sauce. This pudding may be made of peaches, apples, cherries or blackberries; sufficient for 12 persons. For a small family ½ the above quantities will suffice.

363.**Rye Bread Pudding.**— Stir 1 tablespoonful butter to a cream and add by degrees ½ cup sugar and the yolks of 6 eggs; stir this for ½ hour; then add ¼ pound finely pounded almonds, ½ teaspoonful cloves, 1 teaspoonful cinnamon, alittle cardamon and nutmeg and ¼ pound rye bread which has previously been cut into slices, dried in the oven and rolled fine with a rolling pin; add lastly the grated rind of 1 lemon, asmall glass of Cognac or rum and the whites of the 6 eggs, beaten to a stiff froth; butter a pudding form, sprinkle with bread crumbs, fill in the mixture and cover and set the form in a kettle of boiling water; the form should only be immersed in water half way; boil 1½ hours, keeping the kettle closely covered; serve with brandy, wine or hard sauce.

364.**Apple Pudding (German art).** — Pare, core and cut into quarters 6 good sized tart apples, put them in a stewpan with a little water and boil till half done; then carefully remove the apples to a pudding dish, pour 3 tablespoonfuls raspberry syrup or jelly over them and set aside to cool; place a saucepan over the fire with 1 pint milk and ½ tablespoonful butter; as soon as it boils put in 1 cup sifted flour and stir until the mixture forms into a smooth paste and loosens itself from the bottom of saucepan; transfer it to a dish; stir 1 tablespoonful butter to a cream and add alternately 98 the yolks of 5 eggs, 5 tablespoonfuls sugar and the paste, a spoonful at a time; when this is well blended together add the grated rind of 1 lemon, ½ cup finely chopped almonds and lastly the beaten whites of 5 eggs; pour this mixture over the apples and bake in a medium hot oven for ¾ hour; it may be served with wine, fruit or hard sauce or may be dusted with sugar and served without a sauce. NOTE. — When peaches, cherries, plums or berries are used they need not be cooked before baking.

365.**English Apple Pudding.** — Butter a deep pudding dish and sprinkle with bread crumbs; line the sides of dish with a rich pie crust and put a narrow strip around the bottom so as to leave the center of bottom uncovered; next fill the dish with finely cut apples with some sugar sprinkled between them; add a very little nutmeg, apinch of cinnamon and a little butter in small pieces; cover with the same crust and bake 1 hour; when done turn the pudding out onto a dish and serve with hard sauce. (See Sauce.)

366.**Apple Pudding (boiled).** — 1 cup finely chopped suet, 3 cups flour, 1 cup milk, 3 eggs, ½ cup sugar, 3 cups finely cut apples, ½ teaspoonful salt, alittle grated lemon peel and 2 teaspoonfuls baking powder; sift flour, salt and powder into a bowl and add grated lemon peel and suet; next add the yolks of the eggs and mix the whole with the milk to a stiff batter; then add the beaten whites of the eggs; dust the apples with flour and stir them into batter; butter a pudding form, sprinkle with bread crumbs, fill in the mixture, cover tightly and boil 2 hours; serve with hard sauce. NOTE. — The apples and dough may be put in layers in the form, putting first a layer of dough, then a layer of apples, then dough, and so on. This pudding may also be made of apricots, peaches, figs, plums or currants, citron or raisins.

367.**Apple Bread Pudding (German art).**— Pare and cut into slices 8 large tart apples; soak a 5 cent loaf of bread in cold water; when soft press it out and put in a saucepan over the fire with 2 tablespoonfuls butter; stir for 5 minutes and transfer it to a dish to cool; stir 4 tablespoonfuls sugar with the yolks of 4 eggs 99 to a cream and add the bread, the sliced apples and lastly the whites of the eggs beaten to a stiff froth; butter a pudding form, sprinkle with bread crumbs, fill in the bread mixture, boil 2 hours and serve with hard sauce.

368.**Bread Pudding (with Apples).**— Cut 3 slices of bread, ½ inch in thickness, from a 9 cent loaf of bread and soak them in cold water for 10 minutes; press out and put them over the fire in a saucepan with 1 tablespoonful butter; stir for 5 minutes, or until it has formed into a compact mass; transfer it to a dish; when cold stir the yolks of 3 eggs with 2 tablespoonfuls sugar to a cream and add the bread; then add 2 cups finely chopped apples, 2 tablespoonfuls fine bread crumbs and lastly the beaten whites of the eggs; butter a form, sprinkle with bread crumbs, fill in the bread mixture, close tightly and boil for 2 hours; serve with hard, fruit or wine sauce.

369.**Apple Rice Pudding (German art).**— Place a saucepan with ½ pound rice covered with cold water over the fire and boil 5 minutes; drain in colander, rinse off with cold water and return rice to saucepan; add 1 quart milk, ½ teaspoonful salt and boil till tender; pare and cut into slices 6 large tart apples and stew them in 2 tablespoonfuls butter till nearly done; put them into a pudding form; when rice is cold mix it with ¼ pound sugar, the yolks of 6 eggs and lastly the beaten whites of the eggs; pour it over the apples, bake in the oven and serve with the following sauce:—Put the apple peels and cores in a saucepan, cover with water and boil till tender; strain through a jelly bag, return the liquor to saucepan, add the juice of 1 lemon and boil 5 minutes; add 1 cup sugar and let it boil 5 minutes; serve with the pudding.

370.**Apple Pudding à l'allemande.**— Pare and core 6 medium sized greening apples, put them in a long, shallow tin pan, add 2 cups boiling water, cover with another pan of same size and boil 5 minutes; drain off the water and put them into a pudding dish of a size large enough to admit of the apples standing side by side. Pre-

pare the pudding batter as follows:—Put 1 cup milk in a 100 saucepan over the fire, add 1 tablespoonful butter and when it boils add 1
cup sifted flour, stirring constantly; continue the stirring until the
mixture has formed into a smooth paste and loosens itself; then take
it off the fire and let cool; in the meantime stir 1 tablespoonful butter
to a cream and add 4 tablespoonfuls sugar and the yolks of 4 eggs,
adding 1 yolk and 1 spoonful sugar at a time, stirring well; then add
a little of the paste and continue in this way until all the paste, yolks
and sugar are well mixed; add lastly the whites of the 4 eggs, beaten
to a stiff froth; put 1 teaspoonful currant jelly into each apple, pour
the batter over the apples and bake ¾ hour; serve with hard sauce.

371.**Old-Fashioned Apple Pudding.**— ½ pound finely chopped
suet, 1 pound flour, 1 teaspoonful salt and 1 cup cold water; sift
flour and salt into a bowl, add the suet and mix the whole with the
water into a stiff paste; roll it out on a floured board ¼ inch in
thickness, put in the center ½ dozen finely cut tart apples, sprinkle 1
tablespoonful sugar and a little flour between them and add a pinch
of nutmeg and 1 teaspoonful butter in small pieces; dip a large napkin in hot water, ring out and dust it with flour; cover the apples
with the paste, lay the pudding in center of cloth, fold the cloth
together and tie it tightly; have a large kettle of water with ½ tablespoonful salt over the fire; as soon as it boils put in the pudding,
cover the kettle and boil 2 hours; serve with hard, brandy or cherry
wine sauce and if liquor is objected to serve with nutmeg sauce. The
pudding should be served as soon as taken from the water. For a
small family half these quantities will be sufficient.

372.**Roly-Poly.**— 1 cup finely chopped suet, 2 cups prepared
flour, 1 egg and ¾ cup water; mix this into a stiff dough, roll out ⅛
of an inch in thickness, brush it over with beaten egg and sprinkle
over 1 tablespoonful bread crumbs; put on a layer of finely cut apples, sprinkle over 1 spoonful sugar, roll the dough up like a music
sheet, brush the outside all over with beaten eggs and sprinkle with
fine bread crumbs; dip a napkin into hot water, wring out dry and
dust the inside with flour; put the pudding in center of cloth, fold
101 the napkin around it, lap the ends over and fasten with a pin; tie
a string around it, drop into slightly salted boiling water and boil
for 2 hours; serve with the following sauce:—Mix 1 tablespoonful
cornstarch with ½ cup cold water and add 1 cup boiling water and 2

tablespoonfuls butter; boil 5 minutes, strain through a sieve, add 1 cup sugar, alittle lemon juice and 1 cup sherry wine; or serve with hard sauce.

373.**Roly-Poly Tutti Frutti.**— Prepare a dough the same as in foregoing recipe; roll out ⅛ of an inch thick, brush over with beaten egg and sprinkle 1 tablespoonful bread or cracker crumbs over it; pare, core and slice ½ dozen tart apples and put them with 1½ tablespoonfuls butter and 3 tablespoonfuls sugar in a saucepan; add ½ cup currants, the same quantity seedless raisins and finely cut citron; cover saucepan and stew over the fire till apples begin to soften; pour them into a dish and when cold spread the apples over the dough; lay 2 tablespoonfuls currant or apple jelly in small pieces all over the apples; then finish the same as Roly-Poly; serve with the following sauce:—Stir 2 tablespoonfuls butter with 1 cup powdered sugar to a cream and add by degrees 2 whole eggs, alittle nutmeg and 4 tablespoonfuls Jamaica rum or brandy; or serve with lemon or nutmeg sauce.

SOUFLÉS, PANCAKES, OMELETS AND FRITTERS.

374.**Plain Souflé.**— Boil 1½ cups milk with ½ tablespoonful butter and add, stirring constantly, 1½ cups sifted flour; stir till it has formed into a smooth paste and loosens itself from bottom of saucepan; transfer the paste to a dish and set aside to cool; stir 1 tablespoonful butter to a cream and add alternately the yolks of 6 eggs, 6 tablespoonfuls sugar, the grated rind of 1 lemon, the paste and lastly the beaten whites of the eggs; butter a pudding dish, sprinkle with bread crumbs and fill it ½ full of fruit—either peaches, pears pared and cut into quarters, cherries without the pits, currants, raspberries or finely cut apples; blackberries or 102 huckleberries may also be used; sprinkle some zwieback crumbs between the fruit, add sufficient sugar to sweeten, pour over the souflée mixture and bake 1 hour. All kinds of stewed or preserved fruits may be used the same way; serve with claret or fruit sauce.

375.**Almond Souflé.**— Boil 1 cup milk with ½ tablespoonful butter; mix 1 cup rice flour with 1 cup cold milk; stir it into the boiling milk; continue boiling, stirring constantly, until it has formed into a smooth paste and loosens itself from bottom of saucepan; remove the paste from fire, mix it with the yolks of 2 eggs and set aside to cool; stir ½ tablespoonful butter to a cream and add, alternately, 6 tablespoonfuls sugar, the yolks of 6 eggs and the paste; add first 1 spoonful sugar to the butter, then 1 yolk, then a small spoonful paste; stir each part well before another is added; the stirring is best done with a potato masher; when these ingredients are well mixed add by degrees ½ cup finely chopped or grated almonds; add lastly the beaten whites of 6 eggs and fill the mixture into a white porcelain pudding dish which has been well buttered and sprinkled with bread crumbs; bake in a medium hot oven for 40 minutes; when done take the souflée from the oven, dust with powdered sugar, set the dish either in an ornamented silver dish or fold a napkin around it and serve at once with raspberry sauce. NOTE.—Plain flour may be substituted for rice flour; this is sufficient for 10 persons.

376.**Lemon Souflé.**— Boil 1 cup milk or cream with ½ tablespoonful butter; mix 1 cup sifted flour with 1 cup cold milk and stir it into the boiling milk; continue stirring until the contents have formed into a smooth paste and loosens itself from bottom of saucepan; transfer it to a dish and set aside; when cold stir ½ tablespoonful butter to a cream and add, alternately, 6 tablespoonfuls sugar, the yolks of 8 eggs and the paste (by spoonfuls); stir each part well before another is added; then add the grated rind and juice of 1 lemon and lastly the whites of 6 eggs, beaten to a stiff froth; fill the mixture into a well buttered form and bake in a moderately hot oven from 30 to 40 minutes, when done serve at 103 once with wine cream sauce and sprinkle the souflée with powdered sugar.

377.**Vanilla Souflé** is made the same as Almond or Lemon Souflée, omitting the almonds or lemon and adding 2 tablespoonfuls vanilla extract. Extract of lemon may be used the same way.

378.**Orange Souflé** is made the same as lemon, using in place of lemon the juice of 2 oranges and the grated rind of 1. Soufflées may be put into a well buttered form, set in a vessel of hot water and either boiled or baked in the oven.

379.**Chocolate Souflé.**— Boil 4 tablespoonfuls grated chocolate in 1 cup milk; mix 3 tablespoonfuls cornstarch with 1 cup cold milk, stir it into the boiling chocolate, add 1 teaspoonful butter and continue stirring until the contents loosen themselves from bottom of saucepan; transfer the paste to a dish and set aside; when nearly cold stir ½ tablespoonful butter to a cream and add alternately 6 tablespoonfuls sugar, the yolks of 8 eggs and the chocolate paste (by spoonfuls); add lastly the beaten whites of 6 eggs and finish the same as Lemon Souflée; serve with vanilla sauce.

380.**Macaroon Souflé.**— Put into a buttered pudding dish a layer of macaroons and small sponge cakes; over this a layer of cherries from which the pits have been removed; then again a layer of macaroons and sponge cake; continue in this way until the form is filled; whip the yolks of 6 eggs with 1 bottle Rhine wine and 4 tablespoonfuls sugar over the fire to a cream, but do not allow it to come to a boil; pour this over the cake and fruit, set the dish in a vessel of hot water and bake ½ hour; when done draw the souflée to the front of oven; beat the whites of the 6 eggs with 4 tablespoonfuls currant juice or jelly to a stiff froth, spread it over the souflée, set it back in the oven and bake a few minutes; serve without sauce.

381.**Apple Souflé, No. 1.**— Pare and core 6 greening or pippin apples, set in a pan, add 1 quart hot water, cover with another pan of same size and let them steam on top of stove for 5 104 minutes; carefully remove the apples to a pudding dish and set aside to cool; boil 1 cup milk with a little salt and 1 tablespoonful butter and stir in gradually 1 cup sifted flour; continue stirring until the contents have formed into a smooth paste; transfer the paste to a dish and set aside to cool; stir the yolks of 4 eggs with 4 tablespoonfuls sugar to a cream, add by degrees the paste and when well mixed together add the whites of the eggs, beaten to a stiff froth; put a little jelly or marmalade into each apple, pour the mixture over them and bake 1 hour; serve with wine or hard sauce.

382.**Apple Souflé; No. 2.**— Pare 8 or 10 greening or pippin apples, cut into fine slices and put them in a saucepan with 1 tablespoonful butter, ½ cup sugar, 3 tablespoonfuls finely cut citron and the same quantity of seedless raisins and currants; stew this over the fire till apples are tender, but not broken; add 2 tablespoonfuls apple or quince jelly and set aside to cool; if jelly is not handy any kind of marmalade will do; boil 1½ cups milk with ½ tablespoonful butter and stir in 1 cup sifted flour; stir over the fire to a smooth paste; remove from fire and when cold stir 1½ tablespoonfuls butter to a cream and add, alternately, the yolks of 6 eggs, 4 tablespoonfuls sugar, the paste (by spoonfuls) and lastly the whites of the eggs, beaten to a stiff froth; next butter a pudding dish and put in a layer of bread crumbs ⅛ of an inch in thickness; then a layer of apples and little bits of butter; again bread crumbs, again a layer of apples; pour over the top the souflée mixture and bake 1 hour; serve without sauce in the same dish in which the souflée is baked.

383.**Apple Souflé, No. 3.**— Strain 1 quart apple sauce through a sieve, sweeten to taste and add the juice and grated rind of 1 lemon, the yolks of 5 eggs and lastly the whites of the eggs, beaten to a stiff froth; put this into a buttered pudding dish and bake till it cracks on top; sprinkle with sugar and serve without sauce.

384.**Apple Souflé, No. 4.**— Pare, core and quarter 6 apples, cut each quarter into fine slices and put them into a saucepan with 105 1 tablespoonful butter, ½ cup sugar, 2 tablespoonfuls seedless raisins and the same quantity of currants and finely cut citron; cover and stew till apples are tender, but not broken; add 2 tablespoonfuls quince or apple jelly and set aside; when cold butter a pudding dish and sprinkle with bread crumbs; have a souflée mixture prepared the same as in foregoing recipe; put first a layer of the souflée mixture in the dish and sprinkle over some zwieback crumbs; then a layer of the apples; continue in this way with apples and souflée mixture till dish is full; bake 1 hour; when done turn the souflée onto a round dish and serve with fruit or wine sauce; or without sauce and dust with sugar.

385.**Pineapple Souflé.**— Prepare a souflé mixture the same as for Plain Souflée; butter a pudding dish and sprinkle with bread crumbs; put in a layer of the souflée mixture and sprinkle over 1 spoonful zwieback crumbs; put over this a layer of stewed or preserved pineapples cut into small dice, sprinkle over a little zwieback crumbs and cover with souflée mixture; put in another layer of pineapple and a little zwieback crumbs; put the remaining souflée mixture on top and bake 1 hour; when done turn the souflée onto a dish and send raspberry or wine sauce to table with it.

386.**Rhubarb Souflé.**— Pare and cut the rhubarb finely and put it in a saucepan over the fire to boil; add a little water and sufficient sugar to sweeten; when done press it through a sieve; take 1 quart of this stewed rhubarb and mix it with the yolks of 5 eggs and lastly the whites of the eggs, beaten to a stiff froth; bake in buttered dish till it cracks open on top, which will take about ¾ hour; serve without sauce.

387.**Cherry Souflé** is made the same as Pineapple Souflé. Remove the pits from 1 or 2 pounds cherries, sprinkle with sugar and let them stand 1 hour; then put them in alternate layers with the souflée mixture into a well buttered dish and finish the same as Pineapple Souflée; serve with cherry sauce. Peach, apricot and blackberry souflées are made the same way.

106

388.**Gooseberry Souflé.—** Stew the berries with a little white wine, sweeten to taste and finish the same as Rhubarb Souflée.

389.**Raspberry Souflé.—** Press 1 quart raspberries with 1 handful red currants through a sieve, sweeten to taste and mix with the yolks of 6 eggs, 2 tablespoonfuls cream, 3 tablespoonfuls zwieback crumbs and the beaten whites of the eggs; bake ½ hour.

390.**Cherry Omelets.—** Remove the pits from 1 pound cherries, put them with ½ cup sugar and a little water over the fire and stew till done; transfer them to a dish and set aside to cool; mix 2 tablespoonfuls prepared flour with 1 cup milk, the yolks of 6 eggs and lastly the 6 whites beaten to a stiff froth; pour half of this into a hot pan with butter and fry a light brown on the underside; then slip the omelet onto a plate and set it for a few minutes in the hot oven; then take out, put 2 or 3 tablespoonfuls stewed cherries over it, double up and return to the oven until the second one is finished; sprinkle over some sugar and serve with stewed cherries.

391.**Fruit Pancakes.—** Mix 1½ cups sifted flour with ½ teaspoonful baking powder and add ½ teaspoonful salt, the yolks of 3 eggs and 1½ cups milk or water; when this is well mixed stir in the whites of the eggs, beaten to a stiff froth; bake from this mixture 4 large, thin pancakes; wash some ripe strawberries, sweeten with sugar and mash them all up with a silver spoon; put a layer of the mashed fruit over each pancake, lay them on top of one another, dust with powdered sugar and serve.

392. **Huckleberry Pancakes.**— Prepare a batter the same as for Apple Pancakes; put a pan with 1 tablespoonful lard over the fire; when hot pour in some of the batter, about ¼ inch in thickness, and let it bake for a few minutes; then put on a thick layer of huckleberries and sprinkle over 2 tablespoonfuls zwieback crumbs; when done on the underside slip the cake onto a large plate; lay a piece of butter and lard on top of the berries, put over the fryingpan and turn the cake back onto the pan; cover and fry slowly about 6 or 8 minutes; then upset the fryingpan upon a hot dish and 107 sprinkle with sugar; set a plate with the cake over a saucepan of hot water until all are baked in the same manner; lay the cakes on top of one another, dust the whole with sugar and serve. NOTE.—The huckleberries may be stewed with a little lemon juice and sugar and thickened either with zwieback crumbs or cornstarch; aglass of port wine added to it will make a great improvement. They may then be served either separately or put between the cakes. Pancakes with stewed plums or cherries, or any kind of stewed fruit, are very nice.

393. **Strawberry Pancakes.**— Wash 1 quart strawberries and drain them in colander; then prepare 4 large pancakes the same as for Cherry Pancakes; as soon as one is done lay the cake on a plate, cover it with strawberries and sprinkle over some sugar; set the plate over a saucepan of hot water and continue baking until they are all done; lay them over one another with strawberries between and dust the top with fine sugar. Blackberries are treated the same way. Or cover the surface of each pancake with strawberries and sugar, roll each one up separately like a music roll, dust them over with sugar and serve hot.

394. **Cherry Pancakes.**— Remove the pits from 1 pound red cherries; put 1 cup sugar with ½ cup water over the fire and boil a few minutes; put in the cherries and boil 3 minutes; remove from fire and set aside to cool; prepare 4 large pancakes as follows:—Take 2 cups flour, 1 teaspoonful salt, 6 eggs and 1 pint milk; sift flour and salt together, add the milk, the well beaten yolks of the eggs and mix it into a thin batter; beat the batter for 5 minutes with a wooden spoon or German quill; beat the whites of the eggs to a stiff froth and stir them lightly through the mixture; bake 4 large pancakes from this, lay them on one another, with a layer of the stewed cherries between, dust the top with powdered sugar and serve hot. Or

spread over the surface of each pancake a layer of cherries, roll each one up separately, arrange the rolls neatly on a long dish and dust over with powdered sugar; serve while hot. This is a nice dish for dessert.

108

395.**Pancakes (with Currants and Raspberries).** — Strip ½ pound currants of their stems, pick over an equal portion of raspberries, put them in a colander and rinse with cold water; put them in a dish with 1½ cups sugar and let them stand for several hours; bake 3 or 4 medium sized pancakes the same as Cherry Pancake, lay them over one another, with a layer of the sugared fruit between, dust with sugar and serve hot.

396.**Plain German Pancakes.** — 3 cups sifted flour, 2½ cups water, 3 eggs, yolks and whites beaten separately, and 1 teaspoonful salt; put the sifted flour into a bowl, add the salt, make a hollow in the center, add the yolks and mix it gradually with the water into a smooth batter; beat it with a wooden spoon for 5 minutes; then add the whites of the eggs, beaten to a stiff froth; put a large fryingpan with ½ tablespoonful lard and butter over the fire; when hot pour in some of the mixture, sufficient to cover the bottom of pan, about ⅛ of an inch in thickness, shake the pan to and fro and bake till light brown on the underside; slip the pancake onto a large plate, put a little butter and lard in center, put over the fryingpan, turn the pancake back into the pan and bake a light brown; slip the cake onto a hot plate and serve either with syrup, sugar or jelly; continue the baking until all the batter is used.

397.**Lemon Pancakes.** — Bake pancakes the same as in foregoing recipe and when done squeeze over each one some lemon juice, dust with sugar and lay them over one another; stir 1 tablespoonful butter with 3 tablespoonfuls powdered sugar to a cream, set it in a saucepan of hot water and stir till thin; cut the pancakes into pieces, pour some of the sauce over each piece and serve hot.

398.**Peach Pancakes.** — Pare and cut some ripe peaches into fine slices, sprinkle them with sugar and set in a cool place for 1 hour; bake the pancakes the same as in foregoing recipe and lay the peaches between.

399. Apple Pancakes.— Mix 2 cups sifted flour with 2 cups water, ½ teaspoonful salt and the yolks of 3 eggs; when these are well mixed together add the whites of the eggs, beaten to a stiff 109 froth; place a fryingpan with 1 tablespoonful lard over the fire; when hot pour in some of the mixture, about ¼ inch in thickness, put over this a thick layer of very finely cut apples, slip a knife underneath the pancake to keep it from burning and shake the pan too and fro; when the underside is a light brown slip the pancake onto a plate; put a piece of butter and lard on top of the apples, lay the fryingpan over it and turn the pancake over into the pan; cover the pan and let it fry slowly until apples are soft; slip the pancake onto a hot plate and set it over a saucepan of hot water until the remaining mixture is baked the same way. These ingredients will make from 3 to 4 cakes, according to the size of pan. They can be served separately or piled on top of one another. Sprinkle some sugar over each pancake.

400. Apple Fritters.— 1 pint flour sifted with 1 teaspoonful baking powder, 3 eggs, 3 tablespoonfuls sugar, ½ tablespoonful butter, ½ cup milk and 2 cups finely chopped apples; stir butter and sugar to a cream and add the yolks of 3 eggs; then flour and milk, next the chopped apples and lastly the whites of the eggs, beaten to a stiff froth; cut with a spoon a portion from this mixture the size of a large walnut, drop into boiling fat and fry till done; serve dusted with sugar and send wine or snow sauce to table with it. The above recipe will make 20 fritters. If plain flour is used mix it with 1½ teaspoonfuls baking powder and ½ teaspoonful salt.

401. Cherry Fritters.— Remove the pits from 1 pint nice, ripe cherries, mix them with the same ingredients as Apple Fritters, fry in boiling lard, dust with powdered sugar and serve with cherry or wine sauce.

402. Orange Fritters.— Pare and quarter 6 oranges and remove the white skin and pits; mix the orange pieces with the same ingredients as Apple Fritters, drop the mixture, by spoonfuls, into boiling lard and fry a light brown. See that each fritter has 3 pieces of orange and serve with following sauce:—Stir 2 tablespoonfuls butter with 6 tablespoonfuls powdered sugar to a cream and add the yolks of 2 eggs and ½ cup finely cut orange pieces; 110 set the sauce in a saucepan of boiling water and stir till it is melted; then serve. Care should be taken to choose oranges that are not bitter.

403.**Rice Fritters.** — Put 1 cup rice in a saucepan, add cold water and boil 5 minutes; drain in colander and rinse with cold water; return rice to saucepan and add 1 pint milk, ½ teaspoonful salt and ½ tablespoonful butter; boil until rice is thick and soft; transfer it to a dish and when cold mix with 3 tablespoonfuls sugar, the yolks of 4 eggs, 4 tablespoonfuls prepared flour and lastly the whites of the eggs, beaten to a stiff froth; drop this with a tablespoon like small dumplings into boiling lard and fry till done; pile them on a dish, dust over with sugar and serve with snow sauce flavored with wine and a little vanilla.

404.**Cocoanut Fritters.** — Make a batter the same as for Apple Fritters, stir 1 large cup freshly grated cocoanut into it and finish the same as Apple Fritters. Serve with the following sauce: — Boil 1 cup sugar with ½ cup water till it forms a thread between 2 fingers; remove from fire; beat the whites of 2 eggs to a stiff froth; add the boiling hot sugar syrup slowly, beating constantly with an egg beater; then stir in 3 or 4 tablespoonfuls cocoanut.

405.**Currant Fritters.** — 2 cups flour sifted with 1 teaspoonful baking powder, ½ tablespoonful dripping or butter, 2 tablespoonfuls sugar, the grated rind of ½ lemon, ½ cup milk, 2 eggs and ½ cup well washed and dried currants; stir dripping and sugar to a cream and add the yolks of 2 eggs; then the sifted flour and milk; the lemon and currants next; add lastly the whites of the eggs, beaten to a stiff froth; cut with a spoon small portions, the size of a walnut, from the mixture, drop them into boiling lard or dripping and fry a light brown and well done; dust them with sugar and serve with a syrup made as follows: — Boil 1 cup sugar with ½ cup water till it begins to turn yellow; then remove from fire, add a little boiling water, stir for a few minutes and serve. These quantities make 20 fritters.

406.**Walnut Fritters.** — Break the nuts into small pieces and stir 2 cupfuls into a batter made the same as for Apple Fritters. Or 111 bake the fritters plain, prepare a hard sauce, stir some nuts into it and serve with the fritters. Walnut fritters may be served with wine, hard or fruit sauce, or they may be served dusted with sugar without a sauce.

407.**Omelette Souflé à la vanille.** — Stir the yolks of 9 eggs with 3 tablespoonfuls sugar to a cream; add a little salt, 1 teaspoonful vanilla extract and 6 macaroons pounded fine; add lastly the whites of the eggs, beaten to a stiff froth; place an omelet or large fryingpan with butter over the fire; when hot put in ⅓ the egg mixture, shake the pan a little to and fro and bake the omelet to a delicate brown; have ready a buttered dish, turn the omelet into it, with the brown side up, set in the oven and bake another omelet the same way; lay the omelet on top of the one in dish, with brown side up; then bake the third one; lay it on top of the two and bake the whole 10 to 15 minutes; draw them to the front of oven, sprinkle with sugar and hold a red hot shovel over, to brown the sugar; then remove from oven and serve at once. Omelet souflées should be eaten as soon as done.

408.**Omelette Souflé Confitures.** — Prepare 3 or 4 omelets the same as in foregoing recipe, spread over each omelet some peach marmalade or fruit jelly, pour over them when done some warm fruit jelly and serve.

409.**Omelette Souflé (with Chocolate).** — Prepare the omelets the same as in foregoing recipe and sprinkle over each one a tablespoonful grated chocolate.

410.**Omelette Souflé (with Cocoanut).** — Prepare 3 or 4 omelets the same as in foregoing recipe, lay them in a buttered dish on top of one another with thick layers of cocoanut between and bake 10 minutes; dust the souflée with sugar and serve at once.

411.**Rum Or Maraschino Souflé.** — Melt 2 ounces butter in a saucepan, add 2 tablespoonfuls flour and stir for a few minutes; add 1 cup boiling milk and stir till it forms into a smooth paste; remove it from the fire and set aside; when cold stir 2 tablespoonfuls sugar with the yolks of 6 eggs to a cream and add by degrees the 112 paste and 4 tablespoonfuls rum or maraschino; add lastly the beaten whites of the eggs and bake in a well buttered and floured dish ½ hour; serve as soon as baked with lemon cream or wine cream sauce.

412. **Vienna Souflé.** — Place a saucepan with ¾ cup milk, 1 table-spoonful flour and 2 tablespoonfuls butter over the fire and stir till thick; remove from fire and when cold add, alternately, the yolks of 6 eggs, 3 tablespoonfuls sugar, 1 teaspoonful lemon extract and lastly the whites of the 6 eggs, beaten to a stiff froth; fill this into a well buttered and floured dish and bake ½ hour in a medium hot oven; when baked take it from the oven, dust with sugar and serve with raspberry sauce.

413. **Peach Souflé.** — Pare, quarter and stew 1 dozen large, ripe peaches in ½ cup water and 1 cup sugar; when done press them through a sieve and add a little more sugar if not sweet enough; mix with the yolks of 6 eggs and lastly the beaten whites of the eggs; bake in a well buttered dish 40 minutes. Another way is to omit the yolks of the eggs and take only the beaten whites, cherries, huckle-berries and blackberries. Currants and raspberries can also be used the same way.

414. **Apricot Souflé.** — Take a can of California apricots, press them through a sieve, add the syrup and if necessary a little more sugar; mix with 1 cup zwieback crumbs the yolks of 6 eggs and lastly the beaten whites of the eggs; put this into a buttered dish and bake 40 minutes.

415. **Farina Souflé (Vienna art).** — Put 1½ pints milk with 1 table-spoonful butter over the fire; as soon as it boils stir in 6 ounces fari-na; stir over the fire until it has formed into a smooth paste and loosens itself from bottom of saucepan; transfer this paste to a dish; when cold stir ¼ pound butter to a cream, add 4 tablespoonfuls sugar, by degrees the yolks of 8 eggs and the farina paste; add lastly the grated rind of 1 lemon and the whites of the eggs, beaten to a stiff froth; fill the mixture into a buttered pudding dish and bake ¾ hour; serve with fruit or wine sauce.

113

416. **Farina Souflé (Italian art).** — Prepare farina the same as in preceding recipe; when cold stir ¼ pound butter with 5 tablespoon-fuls sugar to a cream; add by degrees the yolks of 9 eggs and the boiled farina; flavor with the rind of 1 lemon; add lastly the whites of the eggs, beaten to a stiff froth; fill a layer, 2 inches thick, into a well buttered pudding dish, spread a thick layer of fruit marmalade

over it and continue with layers of farina mixture and marmalade till all is used; let the last layer be farina; bake ¾ hour and serve with wine cream sauce.

417.**Farina Souflé.** — Boil 1 cup milk with ½ tablespoonful butter; add slowly 4 tablespoonfuls farina and stir till it has formed into a smooth paste and loosens itself from bottom of saucepan; transfer it to a dish and set aside; when nearly cold stir ½ tablespoonful butter to a cream and add alternately 5 tablespoonfuls sugar, the yolks of 5 eggs and the farina paste; stir each part well before another is added; add lastly the whites of the eggs, beaten to a stiff froth, 1 teaspoonful essence of lemon and finish the same as Almond Souflée; serve either with wine cream or fruit sauce.

418.**Strawberry Souflé.** — Wash and press through a sieve 1 quart fresh strawberries; mix them with 6 tablespoonfuls sugar and the beaten whites of 6 eggs; fill this into a buttered dish, sprinkle with sugar and bake slowly 40 minutes; souflées of any kind of fruit jelly or marmalade are made the same way.

419.**Chestnut Souflé.** — Put 30 large chestnuts with cold water over the fire and boil 5 minutes; take them from the fire and remove the outside shells and the brown skins; boil the chestnuts in milk till tender and press them through a sieve; melt ¾ tablespoonful butter, add 1 tablespoonful flour and stir for a few minutes over the fire; add ¾ cup boiling milk, stir and let it boil up, remove from fire and set aside; when cooled off mix it with the chestnut purée and add 4 tablespoonfuls sugar, 1 teaspoonful vanilla and the beaten whites of 6 eggs. This souflée may be baked either in paper boxes or in a dish; dust with sugar when ready to serve.

114

420.**Beignet Souflé.** — Boil ½ pint milk with ½ tablespoonful butter, 1 tablespoonful sugar and add by degrees, while boiling, 1 cup sifted flour; stir constantly till it has formed into a smooth paste and loosens itself from bottom of saucepan; remove the paste from fire and set aside to cool; then mix it with 2 whole eggs and the yolks of 2; place a wide saucepan with lard over the fire, drop with a teaspoon small dumplings into the boiling fat and fry them to a delicate brown; drain them on blotting paper, lay them onto a warm dish, dust with sugar and serve at once.

421.**Potato Souflé.**— Boil 6 large potatoes with the skins in water until done; when cold remove the skins and grate the potatoes on a grater; use only that portion which lies behind the grater and be sure there is 3 cupfuls; then stir 1½ tablespoonfuls butter to a cream and add alternately the yolks of 6 eggs, 6 tablespoonfuls sugar and the grated rind of 1 lemon; add ½ cup ground or pounded almonds and 3 tablespoonfuls dry farina; then add the grated potatoes and lastly the whites of the eggs, beaten to a stiff froth; put this mixture into a well buttered form and bake 1 hour; serve with the following sauce:—Mix 1 tablespoonful butter with 1½ teaspoonfuls corn-starch, add 1 cup boiling water and stir over the fire to a thick, creamy sauce; then add 2 tablespoonfuls sugar, 1 cup sherry wine, alittle lemon juice and ½ teaspoonful vanilla; strain through a sieve and serve.

422.**Vanilla Koch.**— Put 1½ cups milk in a saucepan and add 3 tablespoonfuls sugar, the yolks of 6 eggs and 3 teaspoonfuls flour; mix this well together, place the saucepan in a vessel of boiling wa-ter and stir over the fire till nearly boiling and thick; remove it from the fire and set saucepan in cold water; when cold mix it with 1½ teaspoonfuls vanilla extract and the whites of the 6 eggs, beaten to a stiff froth; fill this into a well buttered and floured mould, set in a pan of hot water and bake in a medium hot oven ½ hour, or till done; when ready to serve send to table either in the same dish or turn onto another dish and send claret or strawberry sauce to table with it. Koch of all kinds should be served immediately upon being done.

115

423.**Cream Koch (boiled).**— Stir together the yolks of 6 eggs with 6 tablespoonfuls cream and 3 tablespoonfuls flour and lastly the whites of the eggs, beaten to a stiff froth; butter a mould, dust with flour, put in the mixture, cover tightly and place in a vessel of boil-ing water; boil slowly 1 hour; or place the form in a pan of hot water and bake in the oven; when baked turn the koch onto a dish and serve with fruit or claret sauce.

424.**Nudel Souflé.**— Boil 1 quart milk with ¼ teaspoonful salt and add 2 cups finely cut home-made nudels; continue the boiling for 15 minutes; then pour the nudels into a dish and when cold stir

1 tablespoonful butter to a cream, add alternately the yolks of 6 eggs and 6 tablespoonfuls sugar; also add the grated rind of 1 lemon; then add the nudels by degrees and lastly the whites beaten to a stiff froth; pour this mixture into a buttered pudding dish, bake 1 hour and serve in the same dish in which it was baked; either set in a silver dish or fold a napkin around it. For sauce boil ½ cup water, dissolve 1 teaspoonful cornstarch in a little cold water and add it to the boiling water; boil a few minutes; then add 1 cup apple or currant jelly; continue boiling, stirring constantly, till jelly is dissolved; then strain through a sieve, add ½ cup white wine and a little sugar if not sweet enough.

425.**Macaroon Souflé.** — ¼ pound macaroons pounded fine, ½ tablespoonful butter, 3 cups boiling milk and 1 tablespoonful flour; put the butter in a saucepan and when melted add the flour; stir for a few minutes; then add the boiling milk and the macaroons; stir this until it forms a smooth paste; transfer it to a dish and set aside to cool; stir 1 tablespoonful butter to a cream and add alternately the yolks of 8 eggs, the macaroon paste and 2 tablespoonfuls sugar; add lastly the whites beaten to a stiff froth; fill this mixture into a well buttered form and bake 1 hour; serve with wine cream sauce or without sauce.

426.**Zwieback Koch.** — Boil 1 pint milk and add 3 ounces rolled zwieback and ½ tablespoonful butter; continue boiling, stirring constantly, until it has formed into a smooth paste; remove 116 from fire and when cold mix with the yolks of 4 eggs, ¼ pound grated hazel nuts, ½ cup sugar and lastly the whites beaten to a stiff froth; fill the mixture into a well buttered dish and bake ½ hour; when done turn the koch out onto a round dish and pour raspberry sauce over it.

427.**Almond Koch (with Snow Sauce).** — Melt 1½ tablespoonfuls butter and add 4 tablespoonfuls sugar and the yolks of 6 eggs; stir this over the fire till thick and smooth; remove from the fire and add ¼ pound finely cut almonds, 1 teaspoonful vanilla extract and the whites of 4 eggs beaten to a stiff froth; put the mixture into a well buttered and floured form, cover, set in a vessel of boiling water and boil 1 hour; when done turn the koch onto a warm dish and pour a snow sauce over it, which is made as follows: — Boil ¾ cup sugar with ½ cup water until it begins to turn yellow; then remove from fire and stir it slowly into the beaten whites of 2 eggs while stirring constantly with an egg beater; flavor with 1 teaspoonful lemon extract; pour the sauce over the turned out koch, set it for a few minutes in the oven and serve; or the sauce may be served separate with the koch.

428.Plain Koch (with Strawberry Chaudeau).— Melt in a saucepan ¼ pound butter and add ¼ pound sugar and the yolks of 8 eggs; stir this over the fire till thick and smooth; remove and mix it with the juice and grated rind of 1 lemon, 1 tablespoonful flour and the whites of 4 eggs beaten to a stiff froth; fill this into a well buttered and floured pudding dish, cover with a tin plate, set dish in a pan of hot water and bake 1 hour; when done turn the koch onto a dish and pour the following strawberry sauce over it:—Beat the whites of 4 eggs to a froth; press the juice from 1 pint strawberries and put it in a saucepan with ½ cup white wine or the juice of 1 lemon, ½ cup sugar and the yolks of 4 eggs; beat this with an egg beater over the fire till it begins to rise; remove instantly, continue beating for a few minutes longer and add the beaten whites; then pour it over the koch or serve it in a sauce dish; or serve the koch with snow sauce.

117

429.Apple Koch; No. 1.— Wash and cut 5 medium sized apples into pieces, put them in a saucepan with a little water and boil till tender; press them through a sieve and mix with 4 tablespoonfuls sugar, the yolks of 4 eggs, alittle grated orange or lemon peel, ½ cup fine bread crumbs and lastly the 4 whites beaten to a stiff froth; bake in a buttered dish 1 hour; when done dust with sugar and serve without sauce.

430.Apple Koch, No. 2.— Pare and cut fine ½ dozen greening or pippin apples, put them in a saucepan with ½ cup white wine, 2 tablespoonfuls sugar and a little lemon or orange peel and let them stew till tender; press through a sieve and set aside to cool; stir the yolks of 4 eggs with 2 tablespoonfuls sugar to a cream and add 2 ounces finely cut citron, 2 ounces grated almonds and 3 tablespoonfuls fine bread crumbs; add the apples and lastly the 4 whites beaten to a stiff froth; fill the mixture into a well buttered form, sprinkle with bread crumbs and bake ¾ hour.

431.Apple Koch (with Almonds and Raisins).— Mix the yolks of 4 eggs with 3 tablespoonfuls sugar and add 4 tablespoonfuls bread crumbs, 1 cup finely cut apples, 2 ounces finely cut almonds, ½ cup seedless raisins, 2 tablespoonfuls cream and lastly the whites beaten to a froth; bake in a buttered dish ¾ hour; when done turn the koch onto a dish; put 1½ cups claret with 3 tablespoonfuls sugar, apiece of cinnamon, alittle lemon peel and a few cloves over the fire; let it boil up, strain and pour over the koch.

432.Jelly Koch.— Stir 2 tablespoonfuls sugar with 1 tablespoonful butter to a cream and add by degrees ½ cup raspberry, currant, apple or quince jelly; continue stirring until well mixed; then add gradually the yolks of 4 eggs and lastly the whites beaten to a stiff froth; bake in a buttered form ¾ hour and serve turned onto a dish dusted with sugar. If this koch is to be boiled take the yolks of 5 eggs and the whites of 2; in serving pour a wine cream sauce around it.

118

433.Cream Koch.— Put in a saucepan 1½ cups milk, the yolks of 6 eggs, ½ cup sugar, 3 teaspoonfuls flour and stir over the fire till nearly boiling; remove it, set saucepan in a pan of cold water and stir till cold; then mix it with the whites beaten to a stiff froth; fill into a well buttered pudding dish, sprinkle over some sugar and finely chopped almonds and bake 20 minutes.

434.Cream Koch (with Sponge Cake).— Spread 8 small slices of sponge cake with quince, apple or currant jelly, put 2 together, cut them through the center and lay into a buttered dish; pour over a little cherry, Madeira or fruit syrup; pour over it a cream koch the same as in foregoing recipe, sprinkle with sugar and bake about 10 minutes. Or dip the cake into the syrup of preserved fruit—either peaches or cherries—and lay some fruit over it; then cover with same cream and bake 15 minutes.

435.Almond Koch.— Stir 1½ tablespoonfuls butter with 4 tablespoonfuls sugar to a cream and add the yolks of 6 eggs and 4 ounces finely chopped blanched almonds; add lastly the beaten whites of 4 eggs and ½ teaspoonful vanilla; butter a small form, sprinkle with flour, put in the above mixture, cover and set the form in a vessel of boiling water; boil gently 1 hour; when done turn the koch onto a dish and serve with strawberry sauce.

436.**Nudel Koch.**— Prepare a nudel dough from the yolks of 2 eggs, apinch of salt and sufficient flour to form a stiff paste; roll out, cut them fine and boil in cream or milk till tender and thick; then set aside to cool; stir 4 tablespoonfuls sugar with 1 tablespoonful butter to a cream and add by degrees the yolks of 5 eggs and the grated rind and juice of 1 lemon; add gradually the nudels and lastly the beaten whites of 3 eggs; put this into a well buttered form and bake ¾ hour; serve with fruit or wine sauce or snow sauce.

437.**Nudel Koch (boiled).**— Butter a pudding form, sprinkle with bread crumbs and lay thin slices of citron all around the form; prepare the nudels the same as in foregoing recipe, add some finely 119 cut citron and put them in the form; boil slowly 1 hour; when done turn the koch onto a dish, sprinkle with sugar, hold a red hot shovel over and pour over the juice of 1 orange; serve with wine cream sauce.

438.**Nut Koch, No. 1.**— Melt 1 tablespoonful butter in a small saucepan and add 3 tablespoonfuls sugar and the yolks of 6 eggs; stir this constantly over a slow fire till thick; remove from fire and when cold mix it with 3 tablespoonfuls almond paste and the whites beaten to a stiff froth; butter a pudding form and dust with flour; set the form in a deep pan of boiling water, cover and set in a medium hot oven to bake 1 hour; serve with the following sauce:—Place a saucepan with 1½ cups white wine, 4 tablespoonfuls sugar and 3 eggs over the fire; beat constantly with an egg beater until it begins to rise; remove instantly, set saucepan for a few minutes in cold water and continue the beating; then pour the sauce in a sauciere and serve. If almond paste is not handy a small cup of almonds grated on a nutmeg grater may be used.

439.**Nut Koch, No. 2.**— Melt 1½ tablespoonfuls butter and add the yolks of 8 eggs and 3 tablespoonfuls sugar; stir this over the fire till thick; remove it from the fire and mix with ¼ pound finely cut hazel or walnuts, 2½ tablespoonfuls fine bread crumbs which have been wet with 3 tablespoonfuls cream or milk and lastly the 8 whites beaten to a stiff froth and mixed with 2 tablespoonfuls powdered sugar; butter a pudding form, sprinkle with bread crumbs, fill in the mixture, cover and set in a vessel of boiling water (the water should reach half way up the form) and boil 1 hour; serve with the following sauce: — Beat the whites of 3 eggs to a froth; let ¾ cup fruit syrup or jelly get boiling hot and add it slowly to the beaten whites, beating constantly with an egg beater; when the pudding or koch is turned out onto a dish pour the sauce around it and serve at once.

440.**Rice Koch.** — Soak 3 tablespoonfuls rice for 2 hours in cold water, drain and dry it on a sieve; then pound it fine and boil in 1 pint cream or milk until thick; when cold stir 1 tablespoonful 120 butter to a cream with 2 tablespoonfuls sugar and add by degrees the yolks of 4 eggs, the rice, 1 tablespoonful rum, the rind of 1 lemon, 2 tablespoonfuls finely cut citron and lastly the whites, which must be beaten to a stiff froth; butter a pudding form, sprinkle with fine bread crumbs, fill it with the rice mixture, close the form tightly and boil 1½ hours; or set the form in a pan of hot water in the oven and bake 1 hour; serve with fruit sauce.

441.**Vanilla Almond Koch.** — Stir 3 tablespoonfuls sugar with the yolks of 5 eggs to a cream and add 2 tablespoonfuls fine bread crumbs, 2 tablespoonfuls almond paste, 1 tablespoonful melted butter, 1 teaspoonful vanilla and lastly the whites beaten to a stiff froth; fill this into a well buttered and floured form, cover, set in a pan of hot water and bake 1 hour; when done turn the koch onto a warm dish, pour over some rum, light it and bring to table in a blaze; send hard sauce to table with it. This koch may also be boiled on top of stove.

442.**Koch (with Orange Chaudeau).** — Melt in a small saucepan 2 ounces butter and add 3 tablespoonfuls sugar and the yolks of 5 eggs; stir this over a slow fire till thick and smooth; remove and mix it with ¼ pound finely grated or pounded nuts, 2 tablespoonfuls finely cut citron, the grated rind of 1 lemon and ½ ounce finely cut candied orange peel; add lastly the whites of 4 eggs beaten to a stiff froth and finish the same as Nut Koch.

443.**Orange Chaudeau.** — Put the juice of 3 oranges and 1 lemon with ½ cup water in a saucepan and add 4 tablespoonfuls sugar and the yolks of 4 eggs; beat this over the fire with an egg beater till nearly boiling; remove, stir for a few minutes longer and serve either in a sauciere or pour it over the turned out koch. If liked a little rum may be added to the chaudeau. It is then called Punch Chaudeau.

444.**Koch (with Nut Cream).** — Melt 2 ounces butter and add the yolks of 5 eggs and 2 tablespoonfuls sugar; stir over the fire till thick; remove and mix it with ¼ teaspoonful cinnamon, apinch of cloves, the grated rind of 1 lemon and 1½ tablespoonfuls 121 bread crumbs wet with 2 tablespoonfuls rum or Cognac; remove the shells from ¼ pound hazel nuts or walnuts and grate the kernels on a nutmeg grater; add them to the above mixture with the whites of 3 eggs beaten to a stiff froth and finish the same as in foregoing recipe; serve with the following cream: — Grate 2 ounces almond, hazel or walnuts on a nutmeg grater, put them into 1 pint boiling cream or milk, cover and let it stand till cold; then strain through a fine sieve; put the milk in a saucepan with the yolks of 5 eggs and 2 tablespoonfuls sugar; stir this with an egg beater over the fire till nearly boiling; remove instantly, continue the stirring for a few minutes longer and either pour it over the koch or serve in a sauciere.

445.**Koch (with Chocolate Beguss).** — Melt 1 tablespoonful butter and add 3 tablespoonfuls sugar and the yolks of 5 eggs; stir this over the fire till thick and smooth; remove and set aside to cool; soak 2 milk rolls without the crust in milk or cream; when soft put them with the milk over the fire and boil and stir till it forms into a smooth paste; remove from fire and when cold mix it with the above egg mixture; add 1 cup grated nuts and lastly the whites beaten to a stiff froth; put the mixture into a well buttered and floured dish; cover and set the dish into a deep pan of hot water, set in a hot oven and bake 1 hour; when done turn the koch onto a round dish and pour the following sauce over it: — Boil ¼ pound grated chocolate with 1½ cups water and ½ cup sugar for 10 minutes; or boil ¾ cup sugar with ½ cup water until it begins to get light brown; take from the fire, let it stand for a few minutes and then pour it over the koch.

446.**Beignets of Buns.** — Take some long baker's buns (ones which are a day or 2 old are the best), cut them into halves, dip each half separately into cold milk and lay them on a dish; mix 1 cup sifted flour with ⅛ teaspoonful salt, the yolks of 2 eggs and 1 cup milk to a smooth, thin batter; add lastly the 2 whites beaten to a stiff froth; put a large fryingpan with 1 tablespoonful lard and butter over the fire; when hot dip each half of bun into the batter, lay them in the pan and fry on both sides to a fine brown 122 color; serve on a long dish; dust with sugar and lay 1 spoonful stewed fruit — such as plums, cherries, apples, huckleberries or stewed gooseberries — or some fruit jelly over each one.

447.**Beignets of Zwieback.** — Lay 1 dozen round zwiebacks on a long dish, pour over some cold milk and let them lay until they begin to get soft; dip each one separately into a batter, the same as Beignets of Buns; fry in ½ lard and ½ butter on both sides to a light brown color, dust with sugar and serve with fruit or snow sauce.

448.**Poor Knights (Arme Ritter).** — Cut a long loaf of bread (2 days old) into slices ¼ inch thick, dip each slice into cold milk, lay them on a dish on top of one another, pour a little milk over the whole and let them lay for 10 minutes; beat up 3 eggs with 3 table-spoonfuls milk, dip each slice into the beaten eggs and then fry in ½ butter and ½ lard in a fryingpan to a light brown on both sides; serve on a hot dish dusted with sugar. Stewed or preserved huckle-berries may be sent to table with it or poured over the bread. Jellies of fruit or marmalade may also be served with it.

449.**Apple Beignets.** — Pare and core with an apple corer ½ dozen large apples, cut them into slices ½ inch in thickness, put them in a dish, sprinkle over some sugar, alittle cinnamon and pour over 1 glass rum; let them lay for 2 hours, tossing them up now and then; shortly before serving wipe dry, dip them in a batter, the same as Beignets of Buns, and fry in boiling lard to a light brown color; serve them piled up on a dish, dusted with sugar, and serve with wine or snow sauce; or send to table without sauce.

450.**Poveison.** — Cut a loaf of French bread which is 2 days old (after the crust has been removed) into slices about ½ inch in thick-ness; stew 1 pound dry prunes with a piece of cinnamon and a little sugar and lemon peel; when done drain them on a sieve, remove the pits and boil the liquor down to ½; chop the plums fine and mix them with the liquor; add a little more sugar; spread this plum marmalade a finger thick on one side of each slice of bread, 123 dip them separately into milk, lay onto a dish and let them lay ½ hour; then dip them into beaten egg and fry in boiling lard; when they are all fried dust them with powdered sugar and a little cinnamon; arrange them on a dish with a napkin under and serve hot. The poveison may also be dipped first in beaten eggs and then in bread crumbs. In place of plums any kind of fruit or marmalade may be taken, but it must be thick.

451.**Poveison of Pineapple.**— Prepare the bread the same as in foregoing recipe, spread one side of the slices with a thick layer of pineapple marmalade and finish the same as in preceding recipe.

452.**Beignets à la Marie-Louise.**— Prepare a biroche dough (as in No. 773), roll it out 1 inch thick, cut into rounds and brush half of them over with beaten eggs; put in the center 1 teaspoonful peach or apricot marmalade, cover them with the remaining rounds and press the edges together; cut them out again with a cutter a little smaller than the first one, let them lay on a floured board with a floured napkin under them and set in a warm place for about 1 hour to rise; shortly before serving fry in boiling lard to a light brown color, lay them on a soft cloth, to absorb the fat, and serve with fruit sauce.

453.**Beignets de creme à la française.**— Beat up 3 whole eggs, the yolks of 6 and add 2 tablespoonfuls sugar, ¾ cup cream or milk and a little vanilla; put this into 6 buttered cups, set them in a pan of water, cover and bake till firm; remove them from oven and when cold turn them out and cut each one into 3 slices; lay them onto a tin pan, cut a round piece out of the center and fill up the hole with warm marmalade; when cold dip the beignets first into pounded macaroons, then in beaten egg, then in fine bread crumbs and fry in boiling lard to a light brown; serve on a napkin dusted with sugar and send either a snow, vanilla or caramel sauce to table with them.

454.**Peach Beignets.**— Pare and cut into halves 1 dozen large peaches, sprinkle over ½ cup sugar and pour a glass of Cognac 124 or white brandy over them; cover and let them stand about 2 hours; shortly before serving lay them in rows upon a clean cloth and press another cloth lightly upon them to absorb the moisture; have ready a batter, dip each one separately into it and fry in boiling lard to a light brown color; lay them onto blotting paper, to absorb the fat, dust with powdered sugar and serve with the following sauce:—Heat the peach syrup to boiling point; beat the whites of 3 eggs to a stiff froth, add slowly the hot syrup, beating constantly, and serve.

455.**Batter for Beignets.**— Mix 1 cup sifted flour with a little salt, the yolks of 2 eggs and 1 cup milk to a smooth, thin batter; beat the 2 whites to a stiff froth; then add the batter slowly to the whites, beating constantly; it is then ready for use.

456.**Pineapple Beignets.**— Pare a small, ripe pineapple, cut into very thin slices and remove the hard part in center with a cutter or apple corer, so they have the shape of rings; dip the rings first into sugar, then in batter and fry in boiling lard; lay them on paper or soft cloth, to absorb the fat, dust with sugar and serve with orange snow sauce made as follows:—Put the juice of 3 oranges with ½ cup sugar over the fire to boil for 5 minutes and add a little grated rind; have the whites of 3 eggs beaten to a stiff froth, add the hot orange syrup slowly, beating constantly, and serve; or the beignets may be served without sauce or brushed over with orange glaze. Oranges may be used instead of pineapples.

457.**Beignets of Nudels.**— Prepare nudels from 2 eggs; cut fine and boil them in 3 cups milk with ½ tablespoonful butter and 2 tablespoonfuls sugar until thick; spread them ¼ inch in thickness onto buttered tins and when cold cut them into rounds with a biscuit cutter or small wineglass; spread 1 side with marmalade or jelly, lay 2 and 2 together and dip them in beaten eggs and fine zwieback or bread crumbs; fry in boiling lard and serve on a napkin dusted with sugar.

458.**Beignets à la polonaise.**— Bake ½ dozen small, thin pancakes, put them on paper and cover each cake with the following 125 cream:—Put in a saucepan 1 tablespoonful flour, 1 cup milk, the yolks of 3 eggs, apinch of salt, 1½ tablespoonfuls sugar, ateaspoonful butter and a little vanilla; stir this over the fire till it begins to boil; remove from the fire and when cold spread it over the pancakes; roll them up, cut into 2 pieces, press the edges together, dip each in egg and bread crumbs and bake in boiling lard; serve on a napkin dusted with sugar and send fruit sauce to table with them. They may also be dipped into batter and then fried.

459.**Pannequets à la royale.**— ½ pound sifted flour, 5 ounces melted butter, 6 eggs, 1½ cups cream or milk, ⅛ teaspoonful salt and 2½ tablespoonfuls sugar; stir the 6 yolks, sugar and salt together and add slowly the melted butter, flour and the lukewarm cream; add lastly the 6 whites beaten to a stiff froth; bake this mixture into small pancakes the size of a saucer, spread them with fruit marmalade or jelly, roll them up, lay them together in squares, sprinkle with sugar and hold a red hot shovel over to glaze; arrange them on a dish in two rows over each other and serve with sabayon of oranges or wine chaudeau. These pancakes may be served with any kind of sweet sauce. Stewed fruit may also be laid in center of dish and the pancakes laid around it.

460.**Pannequets à la vanille.**— Prepare some pancakes the same as in Pannequets Meringués, spread over boiled cream, roll each one up separately, cut them into two pieces, arrange them onto a round dish in a circle, sprinkle over some sugar and pounded macaroons and let them heat through slowly in the oven; serve with vanilla sauce.

461.**Pannequets Meringués.**— Mix ½ cup sifted flour with ½ cup cream and add a pinch of salt, 2 whole eggs, the yolks of 6, 1½ tablespoonfuls sugar, 2 ounces finely pounded macaroons and 1 pint rich, sweet cream; mix all the ingredients well together and bake ½ hour before serving thin pancakes from this in an omelet pan; lay the pancakes on a round dish and spread over each a layer of cream the same as Beignets à la polonaise; put the cakes on top of one another; beat the 6 whites to a stiff froth, mix with ½ cup 126 powdered sugar and spread this meringue all over the cakes; set dish in oven for a few minutes; put little bits of bright jelly on top and serve without sauce.

462.**Plain Omelet.**— 3 eggs, 3 spoonfuls, ¼ teaspoonful salt and a pinch of white pepper; stir yolks, pepper, salt and milk together; beat the whites to a stiff froth and add the above mixture slowly to them, beating constantly; put a large frying or omelet pan over the fire with ½ tablespoonful butter; when hot pour in the omelet mixture; do not stir, but as the eggs set slip a broad-bladed knife under the omelet to keep it from burning on the bottom; when done slip the knife under one side of the omelet and double it over; slip it onto a warm plate and set for 2 minutes in a hot oven; serve at once.

463.**Rum Omelet.**— Prepare an omelet the same as in foregoing recipe; when it comes from the oven dust thickly with granulated sugar; pour 4 tablespoonfuls best rum into a cup, light it with a match and pour while burning over the omelet; serve at once; as a dessert sufficient for 3 persons.

464.**Strawberry Omelet.**— Wash and drain in a colander 1 pint strawberries, put them in a dish with ½ cup sugar and set aside until omelet is made. Ingredients for the omelet:—6 eggs, 1 tablespoonful cornstarch mixed with ¼ teaspoonful baking powder, ½ teaspoonful salt, ½ tablespoonful melted butter and 1 cup milk; stir the yolks, salt, flour, powder and milk together; beat the whites to a stiff froth and add the above mixture slowly to them, stirring constantly; put a large frying or omelet pan with ½ tablespoonful butter over the fire; when hot pour in ½ the omelet mixture; do not stir, but as the eggs set slip a broad-bladed knife under the omelet, to prevent burning on the bottom, and shake the pan to and fro; when the underside is a light brown set pan with omelet for a few minutes in oven; then scatter ½ the strawberries over the surface; slip the broad-bladed knife under one side of omelet and double in two, inclosing the fruit; dust over the top with powdered 127 sugar and let it remain in oven till the next one is baked the same way; then serve at once; sufficient for a family of 6 persons.

465.**Huckleberry Omelet** is made the same way as Strawberry Omelet. Omelets of blackberries, peaches, stewed or preserved fruit, such as cherries, plums, etc., are also made the same way.

466.**Orange Omelet.**— Prepare an omelet the same as for Strawberry Omelet; pare and cut fine 4 oranges, remove pits and white skin, mix the pulp of oranges with sugar and finish the same as Strawberry Omelet.

467.**Jelly Omelet** is made the same way as Strawberry Omelet, using jelly instead of strawberries.

468.**Omelette à la française.**— Break 6 eggs into a kettle, beat them with an egg beater until they foam and add 1 teaspoonful salt and a little pepper; place a large frying or omelet pan with 1 heaping tablespoonful butter over slow fire; as soon as butter is hot pour in the eggs and draw them with a spoon slowly from the side of pan to the center; when nearly thick let it stand for a few minutes without stirring and let it get on the underside a light brown; fold it over from both sides and turn onto a dish with the folded side underneath. Some finely minced chives or parsley may be mixed into the eggs before baking.

CHARLOTTES.

469.**Charlotte of Apples, No. 1.**— Pare and quarter 10 good sized apples (greening or pippin) and cut each quarter into slices; put them in a saucepan with 2 tablespoonfuls butter, 6 tablespoonfuls sugar and the grated rind of ½ orange or 1 lemon; cover and let them stew till apples are soft, but not broken; then add ½ cup currant or apple jelly, ½ cup seedless raisins, the same quantity of currants, 2 tablespoonfuls finely cut citron and a little finely cut candied orange peel; cut a large, stale loaf of bread into thin slices 128 about ⅛ inch thick; cut each slice into rounds with a cake cutter or with the cover of a baking powder box, dip each round piece with one side into melted butter and fit them neatly in the bottom and sides of a 2-quart pudding dish with the buttered side towards the dish; lay the rounds so that they lap over one another; then fill the dish with the apples; cover them with a layer of the round pieces of bread with the buttered sides towards the apples; bake in a medium hot oven 40 minutes; when done turn the charlotte onto a dish, dust with powdered sugar and serve.

470.**Charlotte of Apples, No. 2.**— Cut ½ pound rye bread or pumpernickel into thin slices and dry them in the oven; then roll them fine on a pastry board with a rolling pin and mix with ½ cup melted butter, ½ cup sugar and ½ teaspoonful cinnamon; press this into the bottom and on the sides of a pudding dish in such a way that the inside has a complete lining of bread; fill it with apples prepared the same as in foregoing recipe; cover with a thin layer of bread crumbs and bake 40 minutes; when done turn the charlotte onto a dish, dust with sugar and serve with fruit sauce. Some of the sauce may be poured over the charlotte before sending to table.

471.**Charlotte of Peaches.** — Line a form with bread the same as Charlotte of Apples, No. 1; pare and cut into halves 15 ripe peaches; dissolve 1 pound sugar in 1 cup cold water, place it over the fire and boil 5 minutes; add the peaches and boil about 6 minutes; take them out and set aside to cool; add to the syrup ½ cup apple jelly and boil 10 minutes longer; when cold put the peaches into the form, pour over ½ the syrup, cover neatly with bread and bake 40 minutes; when done turn the charlotte onto a dish, pour the remaining syrup over it and serve at once. Charlottes may be made of all kinds of preserved fruits, such as peaches, cherries, apricots, pears or plums.

472.**Charlotte of Cherries.** — Remove the pits from 2 pounds cherries; dissolve 1 pound sugar in 1 cup cold water and boil 5 minutes; put in the cherries and boil 3 minutes; remove the fruit 129 with a skimmer, boil the syrup a little longer and then set aside; line a form with bread the same as Charlotte of Apples, put in the cherries, pour over a little of the syrup, cover with bread and bake 40 minutes; when done turn onto a dish; add to the remaining syrup 1 glass brandy and pour it over the charlotte.

473.**Charlotte of Currants.** — Remove 1 pound currants from their stems and add 1 pound raspberries; wash and put them with 1½ cups sugar into a dish and let them stand for 1 hour; line a form with bread the same as Charlotte of Apples, put in the fruit, cover with bread and bake 40 minutes; when done turn the charlotte onto a dish, dust with powdered sugar and serve.

474.**Charlotte of Pineapple.** — Line a form with bread the same as Charlotte of Apples, No. 1; pare and cut into fine pieces 1 large pineapple; boil 1 pound sugar with 1 cup water, add the pineapple and boil 20 minutes; remove the fruit with a skimmer, boil the syrup a little longer and then set aside to cool; put the pineapple in the form with 2 tablespoonfuls crab apple jelly laid in small pieces between, pour over a little of the syrup, cover with bread and bake 40 minutes; when done turn the charlotte onto a dish and pour the remaining syrup over it.

475.**Charlotte à la polonaise.** — Cut a large stale sponge cake into slices ½ inch in thickness and pour over each slice a little maraschino or Madeira wine; spread the bottom slice thickly with cream frangipane (see Cream), lay over this another slice, spread again with cream and continue until the cake has its original form again; set the cake onto a dish; beat the whites of 6 eggs to a stiff froth, spread it on thickly over the cake, dust with powdered sugar and set for a few minutes in a cool oven; serve with sabayon sauce made as follows:—Place a saucepan over the fire with ½ cup sugar, ½ bottle Rhine wine, the peel and juice of 1 lemon, ½ teaspoonful cornstarch, 1 whole egg and the yolks of 3; beat this with an egg beater over the fire till nearly boiling; remove instantly, continue the beating for a few minutes longer, pour the sauce into a sauciere and serve with the charlotte.

130

476.**Charlotte Russe, No. 1.**— Cover the bottom of a round form with white paper; split and trim 1 pound lady fingers and fit them neatly in the bottom and sides of form; whip 1 quart cream to a stiff froth and add 5 tablespoonfuls powdered sugar and 1½ teaspoonfuls vanilla extract; fill cream into the form, cover with the cakes laid close together and set on ice till wanted; when ready to serve turn the charlotte russe onto a glass dish, remove the paper and serve.

477.**Charlotte Russe, No. 2.**— Boil 3 cups milk with a pinch of salt, ½ tablespoonful butter and ½ cup sugar; mix 4 tablespoonfuls cornstarch with 1 cup cold milk; stir this into the boiling milk and continue boiling for a few minutes; remove from the fire; beat up the yolks of 4 eggs and mix them with the cornstarch; when nearly cold beat the whites to a stiff froth, stir them lightly through the custard and flavor with 1½ teaspoonfuls vanilla; put a round piece of paper in the bottom of a mould; then line the bottom and sides with lady fingers, fill in the cold custard, lay the cakes closely together on top and set on ice till wanted; when ready to serve turn the charlotte russe onto a glass dish and serve with vanilla sauce.

478.**Charlotte à la russe.**— Cover the bottom of a round form with white paper and line the inside of it with sponge cake; cut the cake for the bottom into 3-cornered pieces, lay them with the points towards the center and let them lap over on one another; cut the pieces for the sides as long as the height of form and about 1½ inches wide; cut them a little slanting on one side towards the top, fit them in firmly close to one another and fill the form with the following cream:—Soak 1 ounce gelatine in a little cold water; place a saucepan with 1 pint milk, the yolks of 6 eggs, 1 small cup sugar and 2 teaspoonfuls vanilla over the fire and stir till nearly boiling; remove from the fire, add the gelatine and stir till cold; when it begins to thicken add 1 pint whipped cream and finish the same as in foregoing recipe. If cream is not handy beat the 6 whites to a stiff froth and add them instead of it.

131

479.**Charlotte à la russe (with preserved or stewed Pears).**— Prepare the milk with gelatine and cream the same as in foregoing recipe; cover the bottom of a round deep dish with preserved or stewed pears, also lay some pears on the side of form, pour in the milk mixture and set on ice till cold; when ready to serve beat the whites to a stiff froth and add 2 tablespoonfuls sugar and ½ teaspoonful vanilla; turn the charlotte onto a dish and spread the beaten whites over it; serve with cold strawberry sauce. This may be made of preserved peaches, apricots or any other kind of fruit in the same manner.

480.**Snow Eggs.**— Beat the whites of 6 eggs to a stiff froth and add 3 tablespoonfuls powdered sugar; put a wide saucepan with milk and 1 tablespoonful sugar over the fire; as soon as the milk boils set with a tablespoon large oval-shaped dumplings from the mixture into the milk, draw saucepan to side of stove, cover and let it stand for a few minutes; then turn them over, let the milk come to a boil again, draw it to the side and cover; after a few minutes transfer the dumplings with a skimmer to a sieve and set aside to cool; prepare a crême française au chocolat or vanille (see Cream) in a plain form; when cold turn the cream onto a dish, lay the snow eggs in a circle around it and serve with vanilla sauce.

RICE PUDDINGS AND DISHES MADE OF RICE FORDESSERT.

481.**To Prepare Rice Flour.**— Pick out all the yellow kernels of a quantity of rice and wash it several times in warm water, rubbing it well between the fingers; drain the water off and pour boiling water over it; let it stand until nearly cold; then pour it into a colander and pour cold water over it; when well drained rub the rice between a towel, spread out on shallow tin pans or on thick brown paper and let it dry in a lukewarm oven; when completely dry pound it to a powder in a wedged wooden mortar; it may then be put away in jars for use; or pound the rice while wet, 132 then dry and rub it through a sieve. Another way is to grind the rice in a coffee mill.

482.**Rice Beignets.** — Wash ½ pound rice, put it with cold water over the fire and boil for a few minutes; drain in a colander and rinse with cold water; return the rice to saucepan and add 5 cups milk, 4 tablespoonfuls sugar, ½ tablespoonful butter and ½ tea- spoonful salt; boil until tender; then mix it with 3 well beaten eggs; spread it evenly a finger thick onto buttered tin pans and set aside; when cold cut the rice into rounds or square pieces, dip them in beaten egg and bread crumbs and fry in boiling lard; serve with wine or fruit sauce. Some finely chopped almonds may be boiled in the rice.

483.**Rice Pears.** — Parboil ¼ pound rice in water, drain in a colan- der and rinse with cold water; return it to the saucepan, add 1 pint milk, alittle salt, 2 tablespoonfuls sugar, 1 tablespoonful butter, alittle grated lemon peel and boil slowly till done; remove the rice from the fire; when cooled off add the yolks of 3 eggs and 1 table- spoonful finely chopped almonds; spread this rice onto a long dish and when cold divide it into equal parts with a spoon the size of an egg; form them into the shape of a pear and press a preserved cher- ry or a little marmalade in the center of each; roll them first into fine bread crumbs, then into the beaten white of egg; then roll again in the bread crumbs and fry them a light brown in boiling lard; put a small piece of cinnamon in the end of each piece, to form the stern, and serve hot with wine sauce.

484.**Rice Beignets (with Chocolate).** — Prepare ½ pound rice the same as in foregoing recipe without the butter and add while hot ¼ pound grated chocolate; spread the mixture evenly a finger thick onto buttered tins and when cold cut it into small rounds with a wineglass; dip them into beaten egg and bread crumbs and fry in boiling lard; dust them with sugar and serve without sauce.

485.**Rice Boiled with Raisins.** — Place a saucepan with 1 cup rice over the fire, cover with cold water and boil for a few 133 minutes;

drain in a colander and rinse with cold water; return the rice to saucepan again, add 1 quart milk, ½ teaspoonful salt, 1 tablespoonful sugar and boil 20 minutes; then add ¼ pound well cleansed seedless raisins and boil till done; serve dusted with sugar; or boil the rice in water instead of milk, add very little sugar, the raisins and a little salt and serve with meat.

486.**Rice Coteletten.**— Boil ¾ pound rice in water 5 minutes, drain in a colander and rinse with cold water; return it to the fire and boil in milk with a little salt till tender and thick; when cold mix it with 2 eggs and 2 ounces melted butter; form the rice into cotelettens, brush over with egg, sprinkle with fine bread crumbs and fry in ½ butter and ½ lard a light brown; dust them with sugar and serve with stewed fruit. If the cotelettes are to be sweet add some sugar, finely chopped almonds, the grated rind of 1 lemon, alittle rum or rose water and fry the same way as above recipe; served with wine, lemon, cream, chocolate or fruit sauce.

487.**Rice à la Creole.**— Parboil ½ pound rice in water and drain and rinse with cold water; boil it in 1½ quarts sweet cream or milk and a little salt; when done sweeten with 2 tablespoonfuls sugar and flavor with 1 teaspoonful vanilla extract; in the meantime put 2 tablespoonfuls sugar in a pan and let it roast to a caramel, stirring constantly; dissolve it with a little water; add this to the rice, put on a round flat dish, sprinkle with sugar and hold a red hot shovel over for a few minutes to glaze it.

488.**Lemon Rice.**— Place a saucepan with ½ cup rice over the fire, boil 5 minutes and drain and rinse with cold water; return the rice to saucepan, add 1 pint milk, ½ teaspoonful salt and boil until tender; when done add 2 tablespoonfuls sugar; put it into a blancmange mould and let it stand till cold; peel a lemon very thin, cut the peel into shreds about ½ inch in length and as fine as possible, put them into a saucepan, cover with water and boil ½ hour, changing the water 3 times; lest they should be bitter pour about 1 teacupful fresh water upon them; squeeze and strain the juice of the lemon, add it with 1 cup sugar to the water and 134 shreds and let it stew gently for ½ hour; then set aside to cool; having turned out the jellied rice into a glass dish, pour the syrup gradually over it, taking care that the little shreds of the peel are equally distributed over the whole.

489.**Rice à la française.** — Put ½ pound well washed rice over the fire with cold water, boil 3 minutes, drain in a colander and rinse with cold water; return the rice to saucepan, add 5 cups milk, ½ teaspoonful salt, ½ tablespoonful butter and boil till nearly done; then add 8 finely pounded bitter macaroons, 2 tablespoonfuls finely cut candied orange peel, ½ cup preserved cherries freed from the pits and cut fine and a little orange water; put the rice when done on a round flat dish, sprinkle with sugar, hold a red hot shovel over to glaze and serve with wine sauce.

490.**Rice Cherry.** — Prepare 1 quart cherry sauce as follows: — Boil 1 pound sour cherries in 1 quart water, crack some of the stones so as to obtain the flavor from the pits and add a piece of cinnamon, the peel of 1 lemon, ½ pint wine and a few pounded bitter almonds; sugar to taste; when done stir it through a fine sieve, return the sauce to saucepan and add ½ pound rice which has been previously well washed and boiled in water with a little salt for 5 minutes; then drain and rinse with cold water; boil the rice in the cherry sauce till tender; remove from the fire, mix it with the yolks of 6 eggs and when cold add the whites beaten to a stiff froth; rinse out a mould with cold water, sprinkle the inside with granulated sugar, fill in the rice and set it on ice for 2 hours; then serve with or without sweet cream or vanilla sauce.

491.**Poor Man's Rice Pudding.** — Place a saucepan with 3 tablespoonfuls rice and cold water over the fire, boil 5 minutes, drain the rice in a sieve and rinse with cold water; put the rice in a pudding dish with 1 quart milk, ½ teaspoonful salt, 3 tablespoonfuls sugar and 1 teaspoonful essence of lemon; bake in a slow oven till the rice is real soft and serve when cold.

492.**Rice Custard Pudding.** — Place in a saucepan 3½ tablespoonfuls rice, cover with cold water, boil 5 minutes, drain in 135 a colander and rinse with cold water; return it to saucepan, add 1 pint milk, ½ teaspoonful salt and boil till tender; when cold mix it with 3 well beaten eggs, 3 tablespoonfuls sugar, 1 pint milk and a teaspoonful of lemon essence; put this into a pudding dish, lay 1 tablespoonful butter in small bits on top of it, grate nutmeg over and bake till the custard is set; ½ cup seedless raisins may be added to the pudding if liked; serve without sauce.

493.Rice Pudding (Minutatim).— Place a saucepan with ¼ pound rice covered with cold water over the fire, let it boil a few minutes, pour into a colander and rinse with cold water; return rice to saucepan, add 1 pint milk, ½ teaspoonful salt, ½ tablespoonful butter and boil slowly till tender; transfer the rice to a dish and when cold mix it with 2 tablespoonfuls sugar, ½ cup milk, the finely chopped peel of 1 lemon, the yolks of 3 eggs and lastly the whites beaten to a stiff froth; butter some small cups or moulds, line them with a few pieces of citron cut into small, thin slices, fill the cups 3 parts full with the rice mixture and bake in the oven; when done turn them onto a long dish, with a little jelly onto each one, and serve with lemon or fruit sauce.

494.Rice Snowballs.— Put 1 cup rice covered with cold water over the fire, boil 5 minutes, drain in a colander and rinse with cold water; return rice to saucepan, add 1 quart milk, alittle butter, 2 tablespoonfuls sugar, ½ teaspoonful salt and boil slowly till tender; add a flavoring of essence of almonds; when the rice is done put into teacups and let it remain till cold; then turn the rice out in a glass dish, pour over a cold custard and on the top of each ball place a small piece of bright colored preserves or jelly. NOTE.—The custard may be served separate in a saucer with the balls, but the flavoring of it should correspond with the rice. Vanilla or lemon essence may be used instead of almond.

495.Cold Rice Flour Pudding.— Let 1½ pints milk come to a boil, add 4 tablespoonfuls rice flour, mix with ½ pint cold milk and stir for 10 minutes over the fire; beat up the yolks of 5 eggs and add them with 2 tablespoonfuls orange water to the rice 136 mixture; add lastly the whites beaten to a stiff froth; rinse out a mould with water; sprinkle with coarse sugar, pour in the pudding and put the form either on ice or in cold water; when cold turn it onto a dish and serve with fruit sauce.

496.Rice Soufleé.— Place a saucepan with 1 cup milk and ½ tablespoonful butter over the fire; when it boils mix 1 cup rice flour with milk, add it to the boiling milk and stir constantly; continue stirring until it forms into a smooth paste and loosens itself from the bottom of saucepan; transfer the rice mixture to a dish to cool; stir 1 tablespoonful butter to a cream and add alternately the yolks of 5 eggs, 5 tablespoonfuls sugar and the rice paste; add lastly the grated rind of 1 lemon and the 5 whites beaten to a stiff froth; pour the whole into a buttered souflée or pudding dish and bake ½ hour in a moderate oven; all souflées should be served immediately on being taken out of the oven, or they will sink and be nothing more than an ordinary pudding; when done take it out, sprinkle powdered sugar over and send the souflée to table in the dish in which it was baked, either with a napkin pinned around or inclosed in a more ornamental dish; serve with wine, cream, raspberry or cherry sauce.

497.**Rice Flour Pudding (colored).** — Place a saucepan with 2 cups milk, 4 tablespoonfuls sugar, the grated rind of 1 lemon and ¼ teaspoonful salt over the fire; as soon as it boils mix 4 tablespoonfuls rice flour with 1 cup milk, stir into the boiling milk and stir constantly about 10 minutes; set the saucepan into a vessel of hot water, to keep warm; place another saucepan over the fire with 3 cups fruit juice, ½ currant and ½ raspberry; when it boils mix 4 tablespoonfuls rice flour with 1 cup claret, add it with 1½ cups sugar to the boiling juice and boil, stirring constantly, till contents form a smooth paste; then take a handsome form, brush the inside with almonds or fine olive oil, sprinkle with coarse sugar and fill the rice mixture in alternate layers with a wet spoon into the form — first the white, then the red, and so on until form is filled; smooth each layer evenly with the wet spoon; in serving turn it out carefully 137 and lay a border of whipped cream around it. This makes a very handsome dish. Half the above quantities will be sufficient for a family of 8 persons.

498.**Rose Rice Pudding.** — Place a saucepan with 1 cup water and 2 cups currant or raspberry juice over the fire, add a piece of cinnamon and when it boils mix 4 tablespoonfuls rice flour with 1 cup water; add it to the boiling water, stirring constantly, and boil about 10 minutes; remove the cinnamon and sweeten with sugar to taste; rinse out a form with water, sprinkle with sugar, pour in the contents and place the form on ice; when cold turn it out on a dish and lay whipped cream flavored with vanilla around it; or serve with vanilla sauce.

499.Cold Rice Pudding (with Almonds).— Put ½ pound rice covered with cold water over the fire, boil a few minutes and drain in a colander; return the rice to saucepan, add 1 quart milk, ½ teaspoonful salt and boil till nearly tender; then add 1 cup finely chopped almonds, 2 tablespoonfuls sugar and boil till done; rinse out a melon-shaped form, sprinkle with coarse sugar, pour in the rice when nearly cold and set it aside to cool; in serving turn it out on a glass dish and serve with cold cream or fruit sauce.

500.Rice Snowballs (with Apples).— Place a saucepan with ½ pound rice covered with cold water over the fire, boil 5 minutes, drain in a colander and rinse with cold water; return rice to saucepan, add ½ teaspoonful salt, 1 quart milk and boil till nearly done; then set aside to cool; pare and core 6 large apples (greenings or pippins are the best) and put into each apple a little butter, sugar and finely chopped lemon peel; take 6 small, square pieces of white muslin, such as is used for apple dumplings, dip into hot water and dust them with flour; spread over each cloth a layer of boiled rice about ½ inch thick, put an apple in the center of each one, fold the cloth with the rice around it so that the apple is covered with the rice, drop them into boiling water and boil ½ hour; serve with wine or lemon sauce.

138

501.Rice kalte Shale (with Wine).— Boil 1 cup sugar with 3 cups water and the peel of 1 lemon for 10 minutes, when cold add 1 bottle white wine, the juice of 2 lemons and ¼ pound rice which has been boiled in 2 waters till tender; place the kalte Shale on ice till wanted.

502.Rice Radetzky.— Put ½ pound rice covered with cold water over the fire, boil 5 minutes, drain in a colander and rinse with cold water; return rice to saucepan, cover with water, add ½ teaspoonful salt and boil till done and thick; in the meantime melt ½ tablespoonful butter and add 4 tablespoonfuls sugar; stir this over the fire till light brown; then add 3 tablespoonfuls boiling water and the juice of 1 orange and 1 lemon; let it boil up, add the rice and let it remain a little longer over the fire; remove from the fire, mix it with 3 tablespoonfuls rum and put it on a glass dish in layers with fruit marmalade between it; let the last layer be rice; cover with the whites of 3 eggs beaten to a froth, sweeten with sugar and flavor with vanilla; sprinkle over some finely chopped almonds, set it for a few minutes in oven, to take color, and serve when cold.

503.**Rice (with Strawberries).** — Parboil ½ pound rice, drain in a colander and rinse with cold water; return the rice to saucepan, cover it with water, add the juice of 2 lemons, the peel of 1 and boil till done, but in such a manner that the kernels remain whole; let it drain on a sieve; then put it in a dish and pour over 1 pint boiling sugar syrup; when cold arrange the rice on a glass dish in the form of a pyramid with layers of sugared strawberries between and decorate the whole with large strawberries.

504.**Rice (with Apples).** — Parboil ½ pound rice in water, drain in a colander and rinse with cold water; then place it over the fire with 1 quart milk, add ½ teaspoonful salt, apiece of cinnamon and boil till tender; transfer the rice to a dish and set it aside to cool; in the meantime pare and cut fine 6 large apples (either greenings, pippins or Baldwin's); put them on a soup plate, sprinkle with sugar, alittle cinnamon, the grated rind of 1 lemon and 139 a pinch of cloves; cover them with another plate and let them stand until the rice is cold; next butter a deep pudding dish and put the apples in the bottom of it; then stir 2 tablespoonfuls butter with 2 tablespoonfuls sugar to a cream and add alternately the rice (by spoonfuls) and the yolks of 5 eggs, alittle cinnamon and the finely chopped peel of 1 lemon; add lastly the whites beaten to a stiff froth; spread the rice evenly over the apples, put small pieces of butter over and then sprinkle over some sugar and finely chopped almonds; bake 1 hour in a medium hot oven; serve with or without wine sauce.

505.**Rice (with Marmalade).** — Boil ½ pound rice the same as in foregoing recipe and press it through a sieve; when cold stir ¼ pound butter with ½ cup sugar to a cream and add by degrees the yolks of 6 eggs, the rice and lastly the whites beaten to a stiff froth; fill this mixture into a well buttered pudding dish with layers of fruit marmalade between the rice (let the last layer be rice) and bake in a medium hot oven ¾ hour; send to table sprinkled with sugar and serve with or without sauce.

506.**Rice (dressed with Sugar and Cinnamon).** — Place a saucepan with 1 cup rice covered with cold water over the fire and boil 5 minutes; pour the rice into a sieve to drain and rinse with fresh water; return it to saucepan, add ½ teaspoonful salt, 1 quart milk and boil slowly till done; if the rice should be too thick add more milk; when done fill the rice into a deep dish, lay small pieces of butter all over the top and sprinkle thickly with sugar into which a little cinnamon has been mixed; serve as a dessert.

507.**Fine Rice Pudding (with Oranges).** — Wash ½ pound rice in several waters, put in a saucepan covered with water and boil 10 minutes; drain in a colander and rinse with water; boil ½ pint white wine with ½ pound sugar, the grated rind and juice of 1 lemon and ½ pint water for 5 minutes, add the rice and boil till tender; when done transfer the rice to a dish to cool; place a saucepan in a vessel of hot water with ½ teaspoonful cornstarch, ½ pint white wine, 3 eggs, 4 tablespoonfuls sugar, the juice of 1 140 lemon and ½ ounce gelatine soaked in cold water; stir this over the fire with an egg beater until just about to boil; remove instantly from the fire and stir for a few minutes longer; then set it aside to cool; rinse a nicely shaped mould with cold water, sprinkle with coarse sugar and set it into cracked ice; put in first a layer of rice, then a layer of oranges which have been peeled, cut into slices and freed of their pits, sprinkle with sugar, pour some of the wine sauce over it and continue alternately with the layers until the form is filled; let it stand in ice over night; in serving turn the rice onto a dish and garnish with sugared oranges.

508.**Fried Rice.** — Place ½ pound washed rice in a saucepan with cold water over the fire and boil 5 minutes; drain in a colander, return rice to saucepan, add 1 quart milk, ½ teaspoonful salt and boil slowly till tender; when done remove from the fire, mix with the yolks of 2 eggs and spread it on a flat dish; when cold cut it into strips 1 inch wide and 2 inches long, brush them over with the beaten whites of eggs, sprinkle with fine bread crumbs or cracker dust and fry in ½ butter and ½ lard to a fine golden color; serve on a hot dish dusted with sugar.

509.**Rice Pudding (baked).** — Soak ½ pound rice for 1 hour in water, drain and boil it in water 15 minutes; pour the rice onto a sieve and after draining it well return to saucepan; add 1 quart milk, ¾ teaspoonful salt and boil till tender; add 3 ounces butter, ¼ pound sugar and the grated rind of 1 lemon; when cold mix it with the yolks of 6 eggs, ½ cup finely chopped almonds and 3 tablespoonfuls finely cut citron; add lastly the whites beaten to a stiff froth; fill the whole into a well buttered pudding dish and bake 1 hour in a medium hot oven; serve with fruit sauce.

510.**Boiled Rice Pudding.**— Place a saucepan with ½ pound rice covered with cold water over the fire and let it boil 5 minutes; then pour it into a colander and rinse with cold water; return the rice to saucepan, add 1 quart milk, apiece of cinnamon, the peel of 1 lemon, 1 teaspoonful salt and boil slowly till tender (but the kernels must be whole); when cold stir ¼ pound butter with 1 cup 141 sugar to a cream and add 1 cup finely chopped almonds, 1 cup stoneless raisins, 1 tablespoonful finely cut preserved orange peel and the yolks of 8 eggs; add the rice and lastly the 8 whites beaten to a stiff froth; fill the mixture into a well buttered pudding form and boil 2 hours; serve with wine cream sauce made as follows:—Place a saucepan over the fire with 1 pint white wine, the yolks of 3 eggs, 1 teaspoonful cornstarch, the peel of 1 lemon and 4 tablespoonfuls sugar; stir this with an egg beater until just about to boil; then remove it instantly from the fire, beat the whites to a stiff froth and mix it with the sauce; or send cherry sauce to table with it (see Sauce). This pudding is sufficient for 12 persons. If a smaller one is desired use half the quantities.

511.**Rice Flour Pudding.**— Take 1 pint milk, 2 small cups rice flour, 3 tablespoonfuls butter, the grated rind of 1 lemon, 8 eggs, whites and yolks beaten separately, 4 tablespoonfuls sugar, 2 tablespoonfuls finely cut citron and ½ teaspoonful salt; put the milk with ½ of the butter over the fire; mix the rice flour with a little cold water or milk, add the salt and when the milk boils stir the flour into it; stir until the contents have formed into a smooth paste and loosens itself from the bottom of saucepan; remove it from the fire and set aside to cool; stir the remaining butter to a cream and add alternately the sugar, the yolks and the rice dough (by a spoonful at a time); add the lemon, the citron and lastly the whites of eggs beaten to a stiff froth; blend all well together; have ready a pudding form well buttered and sprinkled with bread crumbs, fill it with the mixture and boil 2 hours; serve with wine or cream sauce the same as in foregoing recipe; sufficient for a family of 10 persons.

512.Rice Flour Pudding (baked). — Place a saucepan with 1 pint milk and 1 tablespoonful butter over the fire and when it boils stir into the boiling milk sufficient rice flour to make a smooth dough that loosens itself from the bottom of saucepan; then remove it from the fire and set aside to cool; stir 1 tablespoonful butter to a cream and add by degrees the yolks of 5 eggs, 5 tablespoonfuls sugar, the grated rind of 1 lemon, the dough of rice flour and lastly 142 the stiff froth of the 5 eggs; bake ½ hour; serve with wine cream sauce.

513.Rice Pudding (baked). — Place a saucepan with ½ pound rice covered with cold water over the fire, let it boil a few minutes, pour the rice into a colander and rinse with cold water; return it to sauce-pan, add 1 quart milk, ½ teaspoonful salt, 1 tablespoonful butter, apiece of cinnamon, the peel of 1 lemon and boil slowly till tender; when done remove the rice from fire and let it cool a little; stir the yolks of 4 eggs with 3 tablespoonfuls sugar to a cream; mix it with the rice, remove the lemon and cinnamon and add lastly the whites beaten to a stiff froth; fill the mixture into a form which has been well buttered and sprinkled with bread crumbs and bake 1 hour in a medium hot oven; when done turn the pudding onto a round plate and serve with fruit sauce.

514.Rice Pudding (with Peaches). — Cut 12 large peaches in halves and peel and boil them in ½ pound sugar syrup; crack the stones, take out the pits, remove the brown skins and boil the pits with the peaches; when done transfer the fruit to a dish and set aside to cool; put ½ pound rice with cold water over the fire and let it boil up; drain in a sieve and rinse with cold water; return rice to saucepan with 1 pint sweet cream and boil till tender, but in such a way as to leave the kernels remain whole; when cold mix the rice with 5 tablespoonfuls sugar, 1 teaspoonful vanilla extract, ¼ pint maraschino and 1¼ ounces gelatine; put the dish in cracked ice and stir 1 pint whipped cream through it; fill into a form, cover tightly, pack it in cracked ice with rock salt sprinkled between it so that the whole form is completely covered with ice; let it remain thus for 2 hours; in serving dip the form in hot water, turn the pudding onto a round dish, lay the peaches all over it and the pits on top of peach-es; pour the syrup around it and serve. Rice with apples, pears, apricots or any kind of fruit may be made in the same manner.

515.**Rice Pudding à la Palerino.**— Put ½ pound rice with cold water over the fire, boil 3 minutes, pour it into a sieve and 143 rinse with cold water; return the rice to saucepan, add 1 bottle white wine, the juice of 4 oranges and ¼ pound sugar and boil slowly till tender; add 1½ ounces gelatine which has been soaked in cold water and stir both till it begins to thicken; in the meantime cut some pre-served pineapples into small dice; boil 6 ounces seedless raisins in sugar syrup till tender; put them with ½ pint preserved cherries on a sieve to drain; thin ¼ pound apricot marmalade with a little water, add it with the pineapples, raisins and cherries to the rice and lastly stir 1 quart whipped cream through it; fill the mixture into a pud-ding form, pack it in ice and rock salt and let freeze for 4 hours; in serving dip the form in hot water, turn its contents onto a round dish, lay preserved apricots around it and put on top some whipped cream, so as to make this fine dish as tempting as possible.

516.**Rice à la Malte.**— Boil 1 pound rice 5 minutes, drain in a sieve and rinse with cold water; return rice to saucepan, cover with water and boil till tender, but in such a way that the kernels remain whole; pour the rice in a sieve and let the water drain off; rub off the skin of 6 oranges with loaf sugar and pound the sugar fine; then peel the oranges, divide them into quarters, remove the pits and lay the fruit on a dish; sprinkle 6 tablespoonfuls sugar over and set them aside in a cool place; press the juice from 6 oranges; put the rice with 1½ pints white wine in saucepan, add ½ pound sugar and boil till thick; add the orange juice and the juice of 1 lemon, let it boil a few minutes longer, remove from the fire and stir the orange sug-ar into it; rinse out a form with white wine, pour in the rice mixture and when cold put it on ice for 3 hours; in serving turn it out on a round dish and garnish with the orange quarters; boil the juice of 3 oranges with ½ cup sugar and ½ cup water to a syrup; when cold pour some over the rice and serve the rest in a sauciere with the pudding.

517.**Rice Pudding à la Wellington.**— Boil ½ pound rice for 15 minutes in water, drain it in a sieve and rinse with cold water; re-turn rice to saucepan, add ½ bottle Rhine wine, the peel of 1 lemon, the juice of 3 lemons and sweeten with sugar; when 144 done pour it into a border form and set it in a cool place; in serving turn it onto a round dish, decorate with preserved or stewed cherries and pour a little white syrup over the rice; have ready a plombière of frozen raspberry ice in a high pointed form and put it in the center of the rice.

518.Rice Pudding (with Pineapple). — Take a sponge cake (one that was baked in a round form), hollow it out about 1½ inches, pour some Madeira wine over and sprinkle with finely chopped almonds; set it in a cool oven to dry a little; then place it on a round dish with a fine napkin and cut it down to ⅓ its height into 12 pieces; boil ¾ pound rice 5 minutes in water, drain and rinse with cold water; return rice to saucepan and boil it in water till nearly done; pour it onto a sieve and drain off the water; then return it again to saucepan with some pineapple syrup and boil till done; mix a few tablespoonfuls finely cut pineapple with it, fill the rice (half warm) into the sponge cake and decorate with slices of pineapple and preserved red cherries; serve with apricot sauce. Preserved pineapple may be used for this.

519.Rice Pudding (with Apples). — Put ½ pound rice in a saucepan with cold water, let it boil 5 minutes, drain in a colander and rinse with cold water; return rice to saucepan, add 1 pint milk, ½ tablespoonful butter, ½ teaspoonful salt and boil till tender; pour into a dish to cool; stir 2 tablespoonfuls butter to a cream with 4 tablespoonfuls sugar and add by degrees the yolks of 6 eggs and the rice; add 2 tablespoonfuls rum and a finely minced lemon peel; beat the whites to a stiff froth and add them to the mixture; pare and cut into quarters 8 large tart apples; put peels and cores in a saucepan, cover with water, place over the fire and boil till tender; strain through a jelly bag; put the liquor in a saucepan and boil 5 minutes; add 1 cup sugar, the rind and juice of 1 lemon and when it boils put in the apples; boil nearly done; remove them with a skimmer and put into a dish to cool; butter a pudding dish and sprinkle with bread crumbs; put in ½ the rice mixture, lay over this the apples and over them the remaining ½ of rice; lay a few pieces of butter on top, sprinkle with a little sugar and cinnamon 145 and bake 1 hour in a medium hot oven; serve with the syrup. Half the above quantities will be sufficient for a family of 6.

520.Rice Pudding (with Cherries). — Put ½ pound rice in a saucepan, cover with cold water, boil 3 minutes, drain in a colander and rinse with cold water; return rice to saucepan, add 5 cups milk, ½ teaspoonful salt and boil slowly till tender (the kernels must remain whole); transfer it to a dish and when cold stir ¼ pound butter to a cream, add alternately 4 tablespoonfuls sugar, the yolks of 6 eggs, the rice and lastly the whites beaten to a stiff froth; butter a deep pudding dish and sprinkle with bread crumbs; put in a layer of rice mixture, then a layer of cherries which have been previously freed from the pits, stewed with sugar and drained on a sieve; or take preserved cherries, drain off the liquor and put them between the layers of rice till the dish is filled; let the last layer be rice; bake 1 hour in a medium hot oven; when done decorate the top with preserved or stewed cherries and send a cherry sauce to table with it. Rice pudding with apricots or peaches may be made the same way.

521.**Rice Pudding (with Almonds).** — Prepare the rice the same as in foregoing recipe; when cold stir ¼ pound butter to a cream, add ¼ pound sugar, the grated rind of 1 lemon, 1 cup finely chopped or ground almonds, the yolks of 6 eggs, the rice and lastly the whites beaten to a stiff froth; fill the mixture into a well buttered form and bake 1 hour; when done dust it with powdered sugar and serve with fruit or wine sauce.

522.**Lemon Rice Pudding.** — Let ½ cup rice come to a boil, drain in a colander and rinse with cold water; return the rice to saucepan, add 1 cup milk, ¼ teaspoonful salt and boil till tender; when done pour it into a dish and set aside to cool; stir 1 tablespoonful butter to a cream, add the yolks of 4 eggs, 2 tablespoonfuls sugar, then the rice and the grated rind of 1 lemon; whip the juice of 1 lemon with the whites of 4 eggs to a froth, stir into the rice mixture, put it in a well buttered pudding dish and bake ¾ hour; serve with wine sauce.

146

523.**Rice Scallop.** — Place a saucepan with ½ pound rice and cold water over the fire, boil 5 minutes and drain in a colander; return rice to saucepan, add 1 quart milk, ½ teaspoonful salt, asmall piece of butter and boil till rice is soft and thick; pare, core and slice fine 10 large apples, put them in a saucepan with 1 tablespoonful butter, 2 tablespoonfuls finely cut citron, ½ cup well cleansed currants, ½ cup sugar and stew slowly until apples are soft; add ½ cup currant jelly or fruit marmalade and set aside to cool; butter a pudding dish, put in a layer of rice and over it a layer of apples; continue in this way alternately with rice and apples till all are used; let the last layer be rice; put little pieces of butter all over the top of rice and bake ¾ hour; serve either hot or cold.

524.**Cream of Rice Flour.**— Mix in a saucepan 4 tablespoonfuls rice flour with 1 quart milk and add 4 tablespoonfuls sugar, 1 ounce finely cut orange peel, the yolks of 6 eggs and the grated rind of 1 lemon; stir this over the fire until it begins to boil; remove it from the fire, add the whites beaten to a stiff froth and serve it when cold in a glass dish; sprinkle it with sugar, hold a red hot shovel over to brown the sugar and serve without sauce.

525.**Rice kalte** Schale **(with Cream).**— Put ¼ pound rice with cold water over the fire, boil 5 minutes, drain in a colander and rinse with cold water; return rice to saucepan and add 1 quart milk, ½ teaspoonful salt, apiece of cinnamon and a little vanilla and lemon peel; when done remove peel and cinnamon and add the yolks of 4 eggs and 1 quart boiling hot cream; sweeten to taste and serve when cold. NOTE.—Milk may be substituted for cream.

526.**Rice Jelly.**— Mix ¼ pound rice flour with 1 quart milk and add ¼ pound sugar, ½ ounce gelatine which has been soaked for a few minutes in cold water, 6 pounded bitter almonds and a little lemon peel; stir this over the fire constantly and boil 15 minutes; remove lemon peel, fill the mixture into a form which has been well rinsed out with cold water and sprinkled with sugar and 147 set the form on ice to cool; in serving turn the jelly out onto a dish and send fruit syrup to table with it.

527.**Apples with Rice Border.**— Place a saucepan with ½ pound rice covered with cold water over the fire, boil a few minutes, drain the rice in a colander and rinse with cold water; then return it to saucepan, cover with 1 pint milk and boil till tender and thick; then stir through it carefully 1 tablespoonful butter, 3 tablespoonfuls sugar, the grated rind of ½ lemon and the yolks of 2 eggs; next butter a mould (special moulds are made for rice border), fill it with rice and set in a warm place for 10 minutes; pare 12 small tart apples, remove the cores without breaking the fruit and boil them in sugar syrup with 1 tablespoonful apricot marmalade or fruit jelly;

turn the rice border onto a dish and lay little pieces of currant jelly on top of it; pile the apples high up in the center; reduce the syrup by boiling a little longer and pour it over the apples. Pears, peaches, apricots and cherries the same way. Preserved fruit may also be used.

528.**Rice Cream.**— Boil 6 ounces parboiled rice in 3 cups sweet cream and a little salt till tender; when done mix it with ½ ounce gelatine which has been soaked in cold water; when this is well mixed with the rice remove it from the fire, add the well beaten yolks of 6 eggs, 3 or 4 tablespoonfuls sugar and when cold add 1 glass maraschino and the 6 whites beaten to a stiff froth; fill this into a form and place it on ice to get firm; in serving turn the cream onto a glass dish and garnish with macaroons.

529.**Rice Cream (with Chocolate).**— Place 6 ounces well washed rice with cold water over the fire, boil for a few minutes, drain in a colander and rinse with cold water; return rice to saucepan, add 3 cups sweet cream, ¼ teaspoonful salt and boil till tender; then mix with 1 ounce gelatine which has been soaked for 5 minutes in cold water; also add 6 tablespoonfuls sugar, 1 teaspoonful vanilla extract, ¼ pound finely grated chocolate and when cold add 1 pint whipped cream; fill the mixture into a form, bury the form in finely cracked ice with a little rock salt sprinkled 148 between and let it remain 4 hours; in serving dip the form into hot water and turn the cream onto a glass dish. NOTE.—The chocolate may be left out of the cream and a chocolate sauce served with it.

530.**Rice Cream (with Fruit).**— Put 6 ounces well washed rice with cold water over the fire, let it boil a few minutes and drain and rinse it with cold water; return the rice to saucepan, add 3 cups sweet cream, alittle salt and boil till done; when cold mix it with 3 tablespoonfuls sugar, the grated rind of 1 lemon and ½ ounce gelatine soaked in cold water; when nearly cold add 1 pint whipped cream and put the mixture into a form with layers of preserved fruit, such as cherries, peaches, plums, etc., between it; set the form into a pail with layers of cracked ice and rock salt sprinkled between; the form should be entirely covered with ice; set in a cold place for 4 hours; in serving turn the cream out of the form onto a glass dish and garnish with fruit jelly.

531.Rice Croquettes (served with Sauce as Dessert). — Parboil ½ pound rice for a few minutes in water and drain in a colander; return the rice to saucepan with 1 pint milk, 3 tablespoonfuls sugar, ½ tablespoonful butter, ½ teaspoonful salt and boil till tender; when done remove it from the fire and mix with the yolks of 3 or 4 eggs, the grated rind of 1 lemon and set it aside to cool; form the mixture into balls the size of a walnut, dip them into beaten white of eggs, roll them in zwieback crumbs and fry in boiling lard; serve with wine sauce.

532.Brioche Dough. — Put 1 pound flour into a dish, add ½ teaspoonful salt, make a hollow in the center, pour 1 cup warm milk in which 1 yeast cake has been dissolved into the center, mix it with some of the flour and set in a warm place for 1 hour; then stir ¼ pound butter to a cream and add by degrees 2 eggs; mix this with the flour and yeast to a soft dough, cover and let it rise to double its height in a warm place; then work it through once more, roll it out with a rolling pin about ⅛ of an inch in thickness, cut it into rounds with a tumbler and put in the center either force meat, fruit jelly or rice; bend them over so that the edges meet; wet them 149 with beaten egg and press the edges firmly together; set in a warm place to rise; then brush over with beaten egg, dust with bread crumbs and fry in hot lard.

533.Rice Brioche. — Prepare a brioche dough the same as in foregoing recipe with the addition of 2 tablespoonfuls sugar and the grated rind of 1 lemon; while this is rising put 1 cup rice in a saucepan with cold water over the fire, let it come to a boil, drain in a sieve and rinse with cold water; return the rice to saucepan, add 1 pint milk, ½ teaspoonful salt and boil till tender; remove it from the fire and when cold mix the rice with the beaten whites of 3 eggs, 3 tablespoonfuls sugar, 4 tablespoonfuls finely chopped almonds, 2 tablespoonfuls finely cut citron and ¼ teaspoonful cinnamon; when the dough has risen to double its height divide it into 2 equal parts and roll each part out on a floured board to ½ inch in thickness; spread over each part an equal portion of the rice, fold together, press the edges firmly, so that the rice cannot fall out, shape it in the form of a long square and lay in buttered tin pans; brush over with the yolks of the 3 eggs, sprinkle thickly with finely chopped almonds and dust with powdered sugar; set in a warm place and let it rise for ½ hour; then bake in a medium hot oven about ½ hour; if the oven should be too hot cover with paper; serve with the following sauce: — Stir 2 tablespoonfuls butter with 8 tablespoonfuls powdered sugar to a cream and add 2 eggs, alittle nutmeg and 4 tablespoonfuls rum with 1 teaspoonful vanilla; stir until white; put the bowl which holds the mixture into a vessel of boiling water, stir over the fire, add gradually 1 cup boiling water, stir to a creamy sauce and serve. NOTE. — Half these quantities will be sufficient for a family of 6 persons.

534.**Poveison (with Prunes).** — Stew some dried prunes with a stick of cinnamon and lemon peel; sweeten with sugar and set aside to cool; remove the pits and chop the prunes fine; reduce the liquor by boiling it down to ½, add the prunes and mix all together; grate the crust from 3 or 4 milk rolls, cut them into slices about ⅛ of an inch in thickness, spread each slice with the prunes and lay 2 pieces together with the prunes between them; after all the slices 150 have been prepared in this way lay them on a long plate, pour some cold milk over and let them stand for ½ hour; then dip them first in beaten egg and then in fine bread crumbs and fry in boiling lard; serve them dusted with sugar.

535.**Croquettes of Nudels.** — Prepare some finely cut nudels with 1 egg and the necessary flour and boil them in milk with a piece of butter, 2 tablespoonfuls sugar and a pinch of salt; when done transfer them to a dish and mix with the yolks of 3 eggs, the grated rind of ½ lemon and 3 bitter macaroons broken into small pieces; set this mixture on ice and when cold form it into round balls the size of an egg; dip them in the white of egg, then in boiling lard and fry in boiling lard to a fine brown; serve with wine or fruit sauce.

536.**Apple Croquettes.** — Prepare a fine apple sauce, press it through a sieve and add some finely chopped almonds and a little vanilla; bake some thin pancakes of eggs, milk and flour, spread some of the apple sauce over each one, roll them up, turn the ends over and fasten them with egg; dip them first into beaten egg, then into fine zwieback crumbs and fry in boiling lard to a nice golden color; dust them with powdered sugar and serve with a sauce made of apple jelly thinned with vanilla liquor or maraschino.

537.**Nudels (with Jelly).**— Boil 2 cups home-made nudels in salt water 15 minutes; then drain in a colander, put the nudels in a saucepan with 1 tablespoonful butter and toss them over the fire till they are well mixed with the butter; then put them in a glass dish alternately with currant or apple jelly in layers between, cover the top with jelly and serve. Acupful of freshly grated cocoanut laid over the top of this dish and dotted with small bits of jelly is a decided improvement.

538.**Apple Scallop.**— 1 pound bread crumbs, ¾ cup sugar, 10 large greening or pippin apples and ¾ cup butter; pare, quarter and cut the apples into fine slices; put ¼ of the butter into a pudding dish and let it melt; then put in a layer of bread crumbs and over them a thick layer of apples; sprinkle over some sugar 151 and lay little bits of butter over; continue this way alternately with bread crumbs, butter, apples and sugar till all is used; let the last layer be bread with bits of butter on top; cover the dish and bake ¾ hour; shortly before the scallop is done uncover the dish and let it brown a little; serve either hot or cold without sauce.

539.**Nudel Scallop.**— Prepare a nudel dough of 1 egg, ½ teaspoonful butter, 1 tablespoonful water, apinch of salt and sufficient flour to make a stiff dough; divide the dough into 4 parts, roll each part out on a paste board as thinly as possible and let them lay for 15 minutes; then cut each part into long strips 1½ inches in width; lay 4 strips on top of one another and cut them fine, about as thick as a straw; shake the nudels apart and spread them on a board to dry for 1 hour; then put them into boiling salted water, boil 10 minutes and drain in a colander; pare, core and cut into fine slices 10 large greening apples; put the apples in a saucepan with 1 tablespoonful butter and ½ cup sugar, cover and stew gently till they begin to soften; remove them from fire, add ½ cup apple or currant jelly and set aside to cool; butter a pudding dish and put in a layer of nudels; lay little bits of butter over and put over the nudels a layer of apples; then nudels again; continue in this way till all is used; let the last layer be nudels; mix 3 eggs with 1 cup milk, alittle sugar and 1 teaspoonful vanilla; pour it over the nudels and bake ¾ hour; serve either hot or cold. NOTE.—Stewed prunes, peaches or cherries may be used instead of apples. This may be served without sauce.

COLD PUDDINGS MADE WITH MILK.

540.**Cocoanut Custard Pudding, No. 1.**— 2 cups grated cocoanut, 1 cup sugar, 5 eggs and 1 quart milk; beat up the eggs to a froth, add the sugar, stir until melted, then add milk and cocoanut; butter a pudding dish, pour in the mixture and bake till the custard thickens; the best way to ascertain when pudding is done is to place the handle of a teaspoon into the center of the pudding; if it is thick remove instantly and set aside in a cool place; 152 serve when cold in the same dish in which it was baked, with a napkin folded around, or place it in an ornamental dish.

541.**Cocoanut Custard Pudding, No. 2.**— Boil 3 cups milk with 1 cup sugar; dissolve 2 tablespoonfuls cornstarch in 1 cup cold milk and add it to the milk; continue the boiling for a few minutes and remove from fire; beat up the yolks of 4 eggs and after the custard has cooled a little add them to it; when cold beat the whites to a stiff froth and stir them into the custard; butter a pudding dish and put in ½ the custard and a layer of macaroons; then a thick layer of cocoanut on top into which 2 tablespoonfuls sugar have been mixed; bake in the oven to a delicate brown color; serve cold without sauce.

542.**Pudding à la Princess.**— Take 1 pound any kind of cake (sponge, pound, cup or raisin cake), which is several days old, cut it into slices and lay them in a glass dish; put on to each slice 1 teaspoonful currant or apple jelly and pour ½ cup sherry wine and the juice of 1 lemon over; soak 1 ounce gelatine in 1 cup milk 15 minutes; place a saucepan over the fire with 3 cups milk, ½ cup sugar, the yolks of 6 eggs and 1½ teaspoonfuls vanilla extract; stir this over the fire till nearly boiling; then add the gelatine by degrees, stirring constantly, but do not allow it to boil; as soon as gelatine is melted remove it from the fire, set the saucepan in cold water and stir its contents till cold; pour the custard over the cake; beat the 6 whites to a stiff froth and beat into it gradually ½ cup currant, cranberry or apple jelly; spread this meringue over the custard and dot it with little bits of jelly laid on in a pattern. Half the above quantities will be sufficient for a small family.

543.**Cold Sponge Pudding.**— Soak 1 ounce gelatine in 1 cup milk for 15 minutes; place a saucepan over the fire with the yolks of 6 eggs, 3 cups milk, 6 tablespoonfuls sugar, 2 teaspoonfuls vanilla extract and stir with an egg beater till nearly boiling; add by degrees the gelatine, beating constantly; remove from the fire and set aside to cool, stirring it now and then; when cold and 153 beginning to thicken have the whites beaten to a stiff froth and stir them lightly through the cream; rinse out a mould with cold water, sprinkle the inside with sugar, pour in the mixture and set in a cool place to form; in serving loosen the edge on top with your finger and turn the pudding onto a dish; serve with cold fruit or claret wine sauce.

544.**Chocolate Pudding.**— Soak 1 ounce gelatine in 1 cup milk; boil ¼ pound grated chocolate in 1 cup water and add 1 pint milk, the yolks of 3 eggs, 5 tablespoonfuls sugar and the gelatine; beat this with an egg beater over the fire till nearly boiling; remove from the fire and set aside to cool, stirring it now and then; when cold and beginning to thicken add the beaten whites of 6 eggs; rinse a form with cold water, sprinkle with sugar, pour in the mixture and set aside to form; serve with vanilla sauce made of the 3 remaining yolks.

545.**Sago Pudding.**— Boil 1 quart milk with a little salt and while boiling sprinkle in slowly ¼ pound sago; continue the boiling until the sago looks clear and is thick; when done remove the saucepan to side of stove add 4 tablespoonfuls sugar; beat up the yolks of 4 or 5 eggs with a little cold milk, add them to the sago and while hot add the whites beaten to a stiff froth; turn it into a jelly mould which has previously been rinsed with cold water and sprinkled with sugar and place it on ice or in cold water till firm; serve with fruit or vanilla sauce.

546.**Sago Pudding with Almonds** is prepared the same as in preceding recipe. While the sago is boiling add 1 cup finely chopped almonds. Walnuts or hazel nuts may be used the same way.

547.**Sago Pudding (Allemande).**— Boil 1 quart milk with a little salt and while boiling add slowly ¼ pound best sago, ½ cup finely chopped almonds and 4 tablespoonfuls sugar; continue the boiling for 20 minutes longer, then remove the saucepan to side of stove, add the beaten whites of 4 eggs and 1 teaspoonful lemon extract; pour the sago into a well rinsed and sugared jelly mould 154 and set aside to cool; put the 5 yolks with 3 cups milk, 1 teaspoonful cornstarch and 3 tablespoonfuls sugar over the fire and stir until it begins to boil; remove instantly, flavor with vanilla or lemon essence and serve cold with the pudding.

548.**Sago Meringue (with Apples).**— Boil ¼ pound sago in 1 quart milk with a little salt and a little butter; in the meantime pare and core 6 large apples, put them into a long tin pan with 1 quart boiling water, cover them with another pan and boil 5 minutes; transfer the apples to a long-shaped pudding dish and put 1 tea-spoonful jelly into each apple; when the boiled sago is cold mix it with the yolks of 4 eggs and 4 tablespoonfuls sugar; pour this over the apples and bake in the oven; when done draw the pudding to the front of oven; have the 4 whites beaten to a stiff froth and add 2 tablespoonfuls powdered sugar and a little essence of lemon; spread this over the pudding, close the oven and let the pudding bake for a few minutes longer; eat without sauce, either hot or cold.

549.**Sago Cream.**— Put ½ pound sago with boiling water over the fire, let it boil 5 minutes and drain on a sieve; return the sago to saucepan, add 3 pints water and boil slowly 1½ hours; add 1 bottle claret, the juice of 2 lemons and the rind of 1; add sufficient sugar to sweeten and boil ½ hour more; rinse out a jelly mould or small cups with cold water, sprinkle them with sugar, fill them with the sago cream and set in a cool place to get firm; serve cold with whipped cream or vanilla sauce; or take currant juice instead of wine and otherwise prepare the same as above; when cold turn the cream onto a glass dish, lay a border of whipped cream sweetened with sugar and flavored with vanilla around it and serve without sauce.

550.**Rothe Grütze.**— Stew ½ pound currants and ½ pound rasp-berries with ½ pint water about 20 minutes; strain them through a jelly bag; put the juice in a saucepan with the same quantity of wa-ter and add sufficient sugar to sweeten; as soon as it begins to 155 boil sprinkle in slowly some of the best sago, allowing 4 tablespoon-fuls sago to 1 quart liquid; add a piece of cinnamon and boil slowly till sago is clear, which will take about ½ hour; stir it constantly; turn it into cups or jelly moulds; eat when cold with milk or cream.

551.**Milk Pudding.**— Boil 1 quart milk with 6 tablespoonfuls sug-ar, the peel of 1 lemon and add 16 sheets red gelatine which has been soaked for 5 minutes in cold water; stir until the gelatine is dissolved; remove it from fire and add 1 pint Rhine wine; pour it into a jelly mould which has been rinsed with cold water and sprinkled with sugar, set it on ice to get firm and serve with vanilla or lemon custard sauce.

552.Fruit Custard Pudding.— Dip 6 small sponge cakes into the juice of 1 can peaches and lay the cake in a glass dish; lay ½ the peaches over the cake and pour some cold custard over it which is made as follows:—Place a saucepan with 1 quart milk, 4 tablespoonfuls sugar, the yolks of 4 eggs and a pinch of salt over the fire and add 2 tablespoonfuls cornstarch; stir constantly till just about to boil; then remove from the fire and when cold pour it over the cake and peaches; beat the whites to a stiff froth, mix with 2 tablespoonfuls powdered sugar, spread it over the top and serve.

553.Macaroon Meringue.— Place a saucepan with 1 quart milk, the yolks of 4 eggs, 4 tablespoonfuls sugar and 1 tablespoonful cornstarch over the fire and stir constantly till just about to boil; remove instantly and when cold stir 1 cup finely chopped almonds through it; put a layer of macaroons in a glass dish, pour over ½ the custard, put another layer of macaroons and then custard again; beat the whites to a stiff froth, add 1 tablespoonful powdered sugar and spread it over the top.

554.Lemon Custard Pudding.— 1 quart milk, 2 tablespoonfuls cornstarch, 3 eggs, the juice of 2 lemons, the grated rind of 1 and ½ tablespoonful butter; mix the cornstarch with a little cold milk; put the remaining milk over the fire, add the butter and as soon as it begins to boil stir in the cornstarch; boil a few minutes, 156 stirring constantly; remove from fire and when cold mix the eggs with ¾ cup sugar, add the lemon juice and rind, stir this to a cream and add gradually to the cornstarch; when well mixed fill it into a buttered pudding dish and bake till the custard is set; serve cold.

555.Cornstarch Pudding.— 1 quart milk, 4 tablespoonfuls cornstarch, 5 eggs, 4 tablespoonfuls sugar and ¼ teaspoonful salt; mix the cornstarch with a little milk; put the remaining milk with sugar and salt in a saucepan over the fire; as soon as it boils add the cornstarch and let it boil for a few minutes, stirring constantly; when done remove it to side of stove, add the well beaten yolks of the 5 eggs; when well mixed together keeping it hot, and beat the whites to a stiff froth and stir them into the mixture; rinse out a mould with cold water, sprinkle with sugar, pour in the mixture and set either in a cool place or on ice; serve with fruit sauce.

556.**Cornstarch Meringue.**— 1 quart milk, 4 tablespoonfuls cornstarch, 4 tablespoonfuls sugar, 4 eggs and 4 tablespoonfuls fruit jelly; bring the milk to a boil and stir in the cornstarch, which should be previously dissolved in a little cold milk; boil a few minutes, stirring constantly; remove it from the fire and while yet hot stir in the yolks beaten up with the sugar and flavor with 1 teaspoonful vanilla; fill it into a glass dish; beat the whites with 4 tablespoonfuls currant or apple jelly to a froth, spread it over the pudding and serve when cold.

557.**Armor Pudding.**— Boil 1 quart milk with the rind of 1 lemon; dissolve 1 cup cornstarch in 1 cup cold milk and add it slowly to the milk, stirring constantly; add ¾ cup sugar and a little salt and continue the boiling for a few minutes; when done remove the saucepan to side of stove, keeping it hot; beat the whites of 6 eggs to a stiff froth and mix them with the cornstarch; turn it into a jelly mould and serve cold with strawberry or vanilla sauce. The mould should be previously rinsed with cold water and sprinkled with granulated sugar.

157

558.**Red Cream Pudding.**— Boil 1 pint fruit juice and 1 pint water; add 4 tablespoonfuls cornstarch dissolved in cold water, sweeten to taste and continue the boiling for 5 minutes; remove it to side of stove, and while yet hot mix it with the whites of 6 eggs beaten to a stiff froth; turn it into a jelly mould which has been rinsed with water and sprinkled with sugar and serve cold with vanilla sauce.

559.**Floating Island.**— Mix 1 quart milk with the yolks of 4 eggs, 4 tablespoonfuls sugar and ½ tablespoonful cornstarch; stir this over the fire until just about to boil; remove instantly and add 1 teaspoonful vanilla essence; set a glass dish on a wet cloth and pour in the hot custard; beat the whites to a stiff froth, spread it over the custard, sprinkle a little sugar over and cover it up for 20 minutes; then set it on ice or in a cool place; serve cold; or beat the whites with 4 tablespoonfuls fruit jelly to a stiff froth, heap this meringue upon the custard when cold and dot it with bits of jelly laid all over it.

560.**Banana Float.**— Put in a saucepan 1 quart milk, ½ table-spoonful cornstarch, ½ cup sugar and the yolks of 4 eggs; set the saucepan in a vessel of boiling water and stir over the fire till nearly boiling; remove instantly, pour the custard into a dish and add 1 teaspoonful lemon extract; when cold have ½ dozen bananas cut into slices and stir them into the custard; beat the whites to a stiff froth, mix with a little powdered sugar, cover the custard with the meringue, set lady fingers around the edge of dish and serve.

561.**Peach Float.**— Blanch 1 cup almonds, chop them very fine and stir into a custard made the same as in foregoing recipe; pare and cut some ripe peaches into eighths and stir them into the cus-tard; put into a glass dish with meringue on top and garnish with kisses or lady fingers, Orange float is made the same as Banana Float.

562.**Lemon Custard.**— Boil 1 quart milk with ½ cup sugar and while boiling stir in 2½ tablespoonfuls cornstarch previously 158 wet with a little cold milk; stir constantly and boil a few minutes; then remove it from the fire and add 2 teaspoonfuls lemon essence and the well beaten yolks of 4 eggs; turn the custard into a glass dish, beat the whites to a stiff froth, spread it over the top, sprinkle a little sugar over and serve when cold.

563.**Lemon Cream Pudding.**— Place a saucepan over the fire with the rind of 1 lemon and the juice of 3; add 1 cup sugar, 1 cup white wine and the yolks of 4 eggs; stir this until nearly boiling; remove it from the fire and add ¾ ounce gelatine which has been soaked for 10 minutes in cold water; when nearly cold add 1 pint sweet cream beaten to a stiff froth, turn into a jelly mould and set on ice to get firm; serve without sauce. Or put in a saucepan the yolks of 6 eggs, ½ pint white wine, 6 tablespoonfuls sugar, the juice of 3 lemons and the peel of 1; stir this over the fire until it begins to boil; remove it and add 1 ounce gelatine which has been previously soaked in cold water; stir until the gelatine is dissolved; when nearly cold add the whites beaten to a stiff froth; turn into a jelly mould and set on ice to get firm; serve without sauce.

564.**Custard Bread Pudding.**— Mix the yolks of 4 eggs with 1 quart milk, 4 tablespoonfuls sugar and 1 teaspoonful vanilla; pour this into a pudding dish and lay 2 slices of buttered bread on top of the custard; bake until nearly done; beat the whites to a froth, mix with 1 tablespoonful sugar, spread it over the pudding and bake to a light brown color; serve cold.

565.**Custard.** — Beat 5 eggs with ½ cup sugar to a cream and add 1 quart milk and 1 teaspoonful vanilla or lemon essence; pour this into a pudding dish and bake in a medium hot oven till done; to ascertain when the custard is done put the handle of a teaspoon into the center of dish; if the custard is thick and jelly-like, and no milk is to be seen, remove instantly from the oven and serve when cold, or pour the custard into small cups, set them in a long pan of hot water and bake in a medium hot oven till the custard is thick.

159

566.**Apple Custard.** — Pare and core 6 medium sized apples, put them in a pan half filled with boiling water, cover with another pan of same size and let them boil till soft all through, but not broken; transfer them carefully to a glass dish, sprinkle over some sugar and when cold put 1 teaspoonful apple, currant or quince jelly in center of each apple; pour over a cold soft custard. For custard mix 1 pint milk with 3 well beaten eggs and add 2 tablespoonfuls sugar and 1 teaspoonful cornstarch; stir over the fire till nearly boiling, flavor with essence of lemon and set aside to cool.

567.**Pineapple Custard.** — Pare and cut a ripe pineapple into small pieces, taking care not to lose any of the juice; put the fruit with sugar into a glass dish and set on ice; boil 1 pint milk; mix 2 tablespoonfuls cornstarch with 1 cup cold milk, add it with a little salt to the boiling milk and stir and boil for a few minutes; remove from fire and add 2 tablespoonfuls sugar, the yolks of 3 eggs and 1 teaspoonful vanilla extract; mix this well together and set aside; when cold pour the custard over the pineapple; have the whites beaten to a stiff froth, mix a little powdered sugar through it and put on top of custard like a pyramid; place the dish for ½ hour on ice before serving.

568.**Strawberry Custard.** — Wash 1 quart strawberries, drain and put them besprinkled with sugar in a glass dish; pour over a cold custard and finish the same as Pineapple Custard.

569.Peach Custard.— Pare and cut into slices some ripe peaches, sprinkle over some sugar and finish the same as Pineapple Custard. Apricot custard is made the same way.

570.Chocolate Fruit Custard.— Dissolve 3 tablespoonfuls grated sweet chocolate in a little milk and mix it with a custard made the same as Pineapple Custard; when cold pour it over strawberries, pineapples, oranges or any kind of preserved fruit and cover with the whites of 3 or 4 eggs beaten to a stiff froth. If preserved fruit is used the syrup may be used for jelly.

160

571.Fruit Custard (with Cake).— Cut some sponge cakes several days old into square pieces; drain off the liquor from a can of peaches; dip each piece of cake into the liquor and lay them in a glass dish; lay the peaches between and pour a cold custard over; spread over the top the whites of 3 or 4 eggs beaten to a stiff froth, sprinkle over a little sugar and serve. Any kind of stale cake may be used up in this way.

572.Tutti Frutti Custard.— Cut any kind of stale cake into small pieces; put a layer of cornstarch custard into a buttered pudding dish, then a layer of cake; sprinkle over the cake some finely cut citron, raisins and currants; continue in this way until 2 layers of cake and 2 of custard are in the dish; cover the top with a meringue made of the whites of 3 eggs beaten to a stiff froth, add a little powdered sugar and bake in the oven for a few minutes, till meringue is a light brown, and serve. The raisins, currants and citron should be boiled for 15 minutes in a little water before adding them.

573.Apple Custard Pudding.— Pare and core 6 large pippin or greening apples, place them in a long pan with 1 quart boiling water, cover with another pan of same size and stew from 5 to 8 minutes, or until a straw will penetrate through them easily; do not allow them to break; then remove the apples carefully to a pudding dish and put 1 teaspoonful currant jelly into each apple; stir 5 eggs with 4 tablespoonfuls sugar to a cream and add 1 teaspoonful lemon extract and 1 quart milk; pour this over the apples and bake till custard is firm; serve cold in the same dish in which it has been baked.

574.**Peach Meringue.** — Boil 3 cups milk with a pinch of salt, ½ tablespoonful butter and 4 tablespoonfuls sugar; mix 2 tablespoonfuls cornstarch with 1 cup milk, stir it into the boiling milk and continue boiling for a few minutes; then remove from fire and set aside; when cold mix it with the yolks of 4 eggs and 1 teaspoonful lemon extract; pare and cut into halves 8 or 10 large, ripe peaches, lay them in a well buttered pudding dish and sprinkle 3 161 tablespoonfuls sugar over; pour over the cornstarch and bake 20 minutes; draw the dish to front of oven; have the whites beaten to a stiff froth, spread them over the top, sprinkle a little sugar over and bake for 5 minutes longer; serve when warm with fruit sauce and when cold with cream or vanilla sauce. Cherries and pineapples may be used the same way.

575.**Tipsy Parson.** — Take a sponge cake several days old, crumble it up fine, put a layer of it in a glass dish and pour over 1 glass wine; then add ½ cup finely chopped almonds, then a layer of whipped cream; then begin over again, by laying another layer of cake crumbs, almonds and cream; continue in this way till all is used; let the last layer be cream. All kinds of cake crumbs can be used, but sponge cake is the best.

576.**Russian Cream Pudding.** — Rub the skin of 2 lemons and 1 orange on ½ pound loaf sugar; pound the sugar fine, pour the juice of the lemons and orange over it, add 1 ounce gelatine (which should be soaked in a little cold water), 1 pint white wine, 4 whole eggs and the yolks of 8 eggs; put this into a tin pail and set it in a vessel of hot water; stir with an egg beater until just about to boil; remove instantly, add ½ pound sugar and stir until cold; then add 1 quart sweet cream beaten to a froth and ½ pint Madeira wine or rum; fill the cream into a form and set it on ice; in serving turn the cream out and garnish with lady fingers and macaroons.

577.**Tapioca Pudding.** — Wash 3 tablespoonfuls tapioca, put it into a pudding dish with 1 quart milk and let it stand for 1 hour; then set the dish on the side of stove to heat gradually; when the tapioca is soft beat up 3 eggs with 3 tablespoonfuls sugar, stir them into the tapioca and flavor with 1 teaspoonful lemon extract; put small pieces of butter over the top and bake in the oven; serve with or without sauce. Preserved peaches may be sent to table with it.

578.**Tapioca Meringue.** — Soak ½ cup tapioca in 1 quart milk for 2 hours; set the dish on the side of stove and let it heat 162 slowly; stir the yolks of 4 eggs with ½ cup sugar to a cream, add them to the pudding and flavor with 1 teaspoonful vanilla; put 1 tablespoonful butter in small pieces over the top and bake till it begins to thicken; beat the whites to a stiff froth and add 1 tablespoonful powdered sugar and a little lemon extract; draw the pudding to the front of oven, spread the meringue over, set it for a few minutes back in the oven and serve.

579.**Tapioca Pudding (economical).** — Soak 1 cup tapioca in 3 cups water for 3 hours; then put the tapioca into a saucepan, set this into a vessel of hot water, add 1 cup boiling water and boil till tapioca is done (or clear); add ½ teaspoonful salt, the juice of 1 lemon and a little grated rind; sweeten with 5 tablespoonfuls sugar, turn into a jelly mould and serve when cold with the following sauce made of 1 egg, 2 teaspoonfuls cornstarch and 2 cups water; sweeten to taste; add a little butter; stir this over the fire till just about to boil; add a little nutmeg, the juice of 1 lemon and serve when cold with the pudding; or serve with custard sauce.

580.**Tapioca Pudding (with Lemon Sauce).** — Soak 1 cup tapioca in 1 quart milk for 2 or 3 hours; put this in a saucepan and add ½ teaspoonful salt and 2 tablespoonfuls butter; set the saucepan in a vessel of boiling water and boil till tapioca is soft; turn into a jelly mould and set aside to cool; for the sauce pare 1 lemon as thinly as possible and boil the skin for 20 minutes, changing the water 3 times; cut the peel in small strips like straws; place a saucepan with 1 cup sugar, ½ cup water and the lemon peel over the fire and boil for 10 minutes; in serving turn the pudding onto a dish and pour the cold sauce over it. This pudding may be served with vanilla or lemon custard sauce. Tapioca prepared this way may also be put into cups.

581.**Tapioca Pudding (with Apples).** — Soak ½ cup tapioca in cold water for 2 hours, then mix it with ¼ cup sugar, 1 quart milk, 4 eggs and 1 teaspoonful vanilla extract; pare and core 6 pippin or greening apples, put them in a pan with water, cover them with another tin pan and let them boil 5 minutes; remove the apples 163 carefully, put them into a pudding dish, pour the tapioca mixture over them and bake in the oven; serve with hard sauce or send it to table without sauce sprinkled with powdered sugar.

582.**Apple Tapioca Pudding.** — Soak 1 cup tapioca over night in 4 cups water; next morning add about 6 large tart apples, chopped very fine (or more, according to size), and add 1 cup sugar; bake slowly until done; to be eaten either warm or cold with cream. Adelicate dish for invalids.

583.**Farina Custard Pudding.** — Boil 3 tablespoonfuls farina in 1 quart milk with ½ teaspoonful salt; when cold mix it with 4 well beaten eggs, ½ cup sugar and 1½ teaspoonfuls essence of lemon; put into a pudding dish and bake till custard is set; eat hot or cold without sauce. This pudding may also be made with a meringue the same as Cornstarch Meringue.

584.**Rose Pudding.** — Soak 1 ounce gelatine in 1 cup cold water 15 minutes; then add 1 cup boiling water, the yolks of 6 eggs, 6 tablespoonfuls sugar, the peel of 1 lemon and 1 pint white wine; stir this with an egg beater till it nearly boils; then remove instantly, add a little cochineal to color it to a beautiful pink and set aside to cool; when cold and beginning to thicken stir in the whites beaten to a stiff froth; turn into a mould previously rinsed with cold water and sprinkled with sugar and set the form on ice for 2 hours; when ready to serve turn the pudding onto a dish and send vanilla sauce to table with it.

585.**Figaro Pudding.** — Soak 1 ounce gelatine in a little water for 15 minutes; place a saucepan with 1 pint sweet cream, the yolks of 4 eggs and 6 tablespoonfuls sugar over the fire and stir until nearly boiling; add the gelatine and stir till it is dissolved; remove it from the fire and set aside to cool, stirring it now and then; when quite cold and beginning to thicken stir in lightly 1 pint whipped cream and flavor with vanilla; rinse out a mould with cold water, sprinkle with sugar, fill in the cream and set in a cool place to form; serve with cold strawberry sauce.

164

586.**Farina Melusine (with Apples).** — Bring 1 quart milk to a boil, add, stirring constantly, 1 cup farina and stir until it forms into a stiff paste and loosens itself from bottom of saucepan; transfer it to a dish; when cold stir 2 tablespoonfuls butter to a cream and add alternately the yolks of 6 eggs, the farina, 4 tablespoonfuls sugar, the rind of 1 lemon and lastly the beaten whites; pare and core 8 large tart apples, put them in a long pan over the fire, add 1 quart boiling water, cover with another pan of same size and steam them 5 minutes (no longer); then remove carefully, lay them into a long shaped pudding dish, put a teaspoonful jelly into each apple and pour the farina mixture over so the apples are entirely covered; bake in a medium hot oven about ¾ hour and serve with the following sauce:—Stir 2 tablespoonfuls butter with 1 cup powdered sugar to a cream, add the yolks of 2 eggs, 2 tablespoonfuls rum or Cognac, alittle nutmeg and lastly the whites beaten to a stiff froth; sufficient for a family of 8 persons.

587.**Farina Beignets.**— Bring 1 pint milk with 1 tablespoonful butter to a boil and add, stirring constantly, 1 cup farina; continue stirring until it loosens itself from the bottom of saucepan; transfer the farina to a dish and when cold mix it by degrees with the yolks of 3 eggs, the grated rind of 1 lemon and 1 tablespoonful sugar; divide this into equal parts the size of an egg, roll them into oblong shapes, dip them into the beaten whites, roll in fine bread crumbs and fry in boiling lard; serve them dusted with sugar or send fruit sauce to table with them.

588.**Farina Pudding (without Eggs).**— Boil 3 pints milk with a little salt and while boiling sprinkle in slowly 5 tablespoonfuls farina; set the saucepan with the farina in a vessel of hot water and continue the boiling for ¾ hour; pour it into a jelly mould which has been rinsed with cold water and sprinkled with sugar and serve cold with either wine fruit or vanilla sauce; or sprinkle with sugar. This pudding may also be eaten with cream or milk, and is also nice cut into slices and fried; or dip the slices into beaten egg, roll in fine bread crumbs, fry and serve for breakfast with meat.

165

589.**Farina Koch (with Chocolate).**— Boil 3 cups milk, add 5 tablespoonfuls farina and stir constantly until thick; transfer to a dish and when cold stir 2 tablespoonfuls butter to a cream, add by degrees the yolks of 5 eggs, 3 tablespoonfuls sugar and add the farina by a spoonful at a time; add lastly the whites beaten to a stiff froth;

divide the mixture into 2 halves; mix ½ with 2 ounces grated vanilla chocolate; put first a layer of the white farina mixture, then a layer of the chocolate mixture into a well buttered pudding dish; repeat this operation, lay on top some almonds cut into strips, sprinkle over some sugar and bake ¾ hour in a medium hot oven; it may be served with or without sauce; send to table in the same dish in which it was baked; either set in a silver dish or pin a napkin around it.

590.**Farina Souflée (with Almonds and Raisins).**— Boil ¼ pound farina in 2½ cups milk and 1½ tablespoonfuls butter until thick; when cold mix it with the yolks of 6 eggs, 3 tablespoonfuls sugar, alittle salt, 2 ounces finely chopped almonds, 10 bitter almonds, 2 ounces seedless raisins and the 6 whites beaten to a stiff froth; bake in a buttered form and serve as soon as done with raspberry or wine sauce.

591.**Farina Pudding (with Almonds).**— Boil 1 quart milk and while boiling sprinkle in slowly 4 tablespoonfuls farina; add 4 table-spoonfuls finely chopped almonds and continue boiling for 20 minutes; sweeten with 4 tablespoonfuls sugar; when done remove from fire and add the yolks of 4 eggs; pour it into a glass dish, beat the whites with 4 tablespoonfuls currant, apple or cranberry jelly to a stiff froth and spread it over the pudding; serve with claret sauce made as follows:—Put ½ pint water with 3 slices of lemon, apiece of cinnamon and 2 cloves over the fire and boil 5 minutes; then add 1 teaspoonful Bermuda arrowroot and boil 2 minutes; take it from the fire, add ½ pint claret, sweeten with sugar, strain and serve with the pudding when cold. If arrowroot is not handy use cornstarch.

592.**Fine Farina Pudding (boiled).**— Boil 1 cup farina in 1 pint milk with a little salt and ½ tablespoonful butter until it 166 be-comes thick and loosens itself from bottom of saucepan; when cold stir ¼ pound butter to a cream and add alternately 5 tablespoonfuls sugar, the yolks of 8 eggs and the boiled farina by a spoonful at a time; add lastly the beaten whites and grated rind and juice of 1 lemon; butter a pudding form, sprinkle with fine bread crumbs, fill in the mixture, close tightly and boil 2 hours; serve with wine cream sauce (see Sauce). NOTE.—This pudding should be served as soon as taken out of the form.

593. Farina Souflée. — Bring 1 pint milk with 1 tablespoonful butter to a boil and add by degrees, stirring constantly, 1 cup farina; continue stirring until it has formed into a stiff paste and loosens itself from bottom of saucepan; then transfer it to a dish and set aside to cool; stir 1 tablespoonful butter to a cream and add alternately the yolks of 5 eggs, 4 tablespoonfuls sugar, the grated rind of 1 lemon and the farina paste by a spoonful at a time; stir with a potato masher until all is well mixed; add lastly the whites beaten to a stiff froth; fill the mixture into a well buttered pudding form and bake ¾ hour; dust the souflée with sugar and serve as soon as done; send raspberry or any kind of fruit sauce to table with it.

594. Farina Mush. — Put 1 quart milk in a saucepan over the fire and when it boils add gradually, stirring constantly, 1 cup farina; add ½ teaspoonful salt, asmall piece of butter and continue stirring and boiling for 10 minutes; then add by degrees 1 pint milk and boil a few minutes longer; serve on a dish dusted with sugar and if the flavor is liked sprinkle a little cinnamon over; some finely chopped almonds may be added if liked; or put the farina into a dish, sprinkle thickly with sugar and hold a red hot poker over it to brown the sugar.

595. Farina Pudding (cold). — Boil 1 quart milk with ½ teaspoonful salt, ½ tablespoonful butter and while stirring constantly sprinkle in slowly 5 tablespoonfuls farina; continue the boiling for 20 minutes; when done remove it to side of stove, add 4 tablespoonfuls sugar, the yolks of 4 eggs, and while hot add the 167 whites beaten to a stiff froth; fill this into a form, set it in a cool place and serve with strawberry sauce made as follows: — Place a saucepan over the fire with 1 cup water; dissolve 1 heaping teaspoonful cornstarch in a little cold water, add it to the contents of saucepan, boil for a few minutes, transfer to a dish and mix with the juice of ½ lemon, alittle Rhine wine and ½ pint fresh strawberry juice.

596. Fine Farina Pudding (with Vanilla Sauce). — Boil 5 tablespoonfuls farina in 1 quart milk with a little salt the same way as in preceding recipe; as soon as done add 4 tablespoonfuls sugar, and while hot mix it with the whites of 6 eggs beaten to a stiff froth; fill the mixture into a jelly mould which has been rinsed with cold water and sprinkled with granulated sugar and set on ice to cool; put the 6 yolks with 1 quart milk and 4 tablespoonfuls sugar over the fire and stir until just about to boil; remove instantly from the fire and flavor with essence of vanilla.

597.**Figaro Pudding.**— Boil 1 pint milk with a little salt and 1 tablespoonful butter and while boiling sprinkle in slowly 8 tablespoonfuls farina; stir and boil till the farina is thick and loosens itself from bottom of saucepan; remove it to a dish to cool; stir 1 tablespoonful butter to a cream and add alternately the yolks of 8 eggs, 6 tablespoonfuls sugar, the grated rind of 1 lemon and the boiled farina by a spoonful at a time; add lastly the whites beaten to a stiff froth; now divide this mixture into 3 parts; color first part by stirring a few spoonfuls cocoa into it; add to second part a little cochineal for the red; the third part leave white; put this into a well buttered form in 3 layers and boil 2 hours; serve with wine cream sauce.

598.**Apples au beurre.**— Pare, cut and quarter 12 large tart apples, remove the cores and put the apples in a wide kettle with ¼ pound melted butter, ¼ pound sugar and 1 teaspoonful vanilla extract; cover and let them simmer over a slow fire for 10 minutes; then turn each piece over, add 2 tablespoonfuls water and let them stew till tender, but not broken; transfer the apples to a dish; pile 168 them up high in center; add 4 tablespoonfuls apple jelly to the syrup, let it boil up and if too thick add a little more water; pour it over the apples and lay around the edge fleurons of puff paste.

599.**Apple Meringue.**— Pare, core and cut into quarters 12 large tart apples and stew them the same as in foregoing recipe; put the cores and peels covered with water over the fire and boil to a pulp; strain through a jelly bag; measure the liquor; allow for 1 pint liquor 1 pound sugar; boil the liquor 10 minutes; then add the sugar; stir until melted; then remove from fire; pile the apples up high in a dish, pour over the jelly and when cold cover with a thick layer of meringue; set it for a few minutes in a cool oven and serve when cold. Apple marmalade may be used instead of apple jelly and put into the dish in alternate layers with the apples.

600.**Apples (with Whipped Cream).**— Pare, core and cut into quarters 1 dozen large tart apples; boil them in rich sugar syrup till tender, but not broken; remove the apples carefully to a glass dish;

boil the syrup a little longer and pour it over the apples; when cold put over a thin layer of currant jelly and over this a thick layer of whipped cream sweetened with sugar and flavored with vanilla.

601.**Peches à la Condé.** — Boil 10 ounces rice for 5 minutes in cold water, drain in a colander and rinse with cold water; return the rice to saucepan with 1½ pints sweet cream and boil until tender and thick; then add 1½ tablespoonfuls butter, 4 tablespoonfuls sugar, apinch of salt, 2 teaspoonfuls vanilla extract and the yolks of 4 eggs; butter a plain border mould, dust well with flour, put in the rice and bake 15 minutes; then set it in a warm place; pare and cut into halves 1½ dozen peaches and boil them in sugar syrup till a straw will pierce through them easily; take the fruit out carefully and lay it on a sieve to drain; crack the pits, take out the kernels, scald them in boiling water, remove the brown skins and cut the kernels into strips; put them into the peach syrup and boil slowly till syrup begins to thicken; put the peaches back into the syrup and set them in a warm place; shortly before serving 169 turn the rice border onto a round dish, put the peaches in center, pile them up, pour the syrup over the fruit and a little over the border and serve at once. Apricots, apples or pears the same way.

602.**Apricots à la parisienne.** — Pare 12 or 15 ripe apricots, cut them in halves and boil in sugar syrup; take them out and set aside to cool; boil 3 cups milk with ½ tablespoonful butter, 4 tablespoonfuls sugar and a pinch of salt; mix 4 tablespoonfuls cornstarch with 1 cup cold milk, stir it into the boiling milk and continue stirring and boiling for a few minutes; remove to side of stove, add the yolks of 4 eggs and 2 teaspoonfuls vanilla extract; and while hot stir in the 4 whites beaten to a stiff froth; rinse out a border mould with cold water, sprinkle thickly with granulated sugar, put in the cornstarch mixture and set on ice to cool; shortly before serving turn the border onto a round dish and pile the apricots high up in the center; reduce the syrup by boiling it a little longer and pour over the fruit and border when cold. Made of apples, peaches, pears, cherries, pineapples, currants or raspberries the same way. NOTE. — Stewed cherries or finely cut pineapples may be stirred into the cornstarch before putting it into the form and the border. When turned on the dish it may be decorated with fresh strawberries, cherries or raspberries. The cornstarch may be put into a round form. When turned out onto a dish lay the fruit all over it, cover the whole with a meringue and serve the syrup the fruit was boiled in as a sauce with it; or serve vanilla sauce with it.

603.**Pineapple Croutes.**— Pare and cut a small pineapple into small pieces and boil them in sugar syrup till tender; take the pineapple out and reduce the syrup by boiling it down; cut a loaf of stale bread into slices ¼ inch in thickness, cut the slices into rounds with a cake cutter and toast them to a handsome brown color; lay them in a pan and dust with sugar; set them for a few minutes in a hot oven to glaze; spread a thick layer of apple marmalade over the bottom of a round dish; also spread each piece of toast with the marmalade; set them around the edge of dish, pour 170 over some of the syrup and put the pineapple in the center. Brioche when a few days old may be used instead of bread.

604.**Cherry Croutes** are made the same as Pineapple Croûtes. Peach Croûtes and Apricot Croûtes are also made the same as Pineapple Croûtes.

605.**Apples in Jelly, No. 1.**— Pare 1 dozen Spitzenberg apples and remove the cores with an apple corer without breaking the fruit; put a wide kettle or saucepan over the fire with sufficient water to cover the apples, let it come to a boil, put in the apples and boil till a straw will easily pierce through them; then transfer them carefully to a long glass dish; boil the liquor down to 1 quart and add 1 cup sugar and 1 ounce gelatine soaked for 15 minutes in a little cold water; stir it into the apple syrup, boil for a few minutes and then set aside; when nearly cold pour the syrup over the apples and set on ice to get firm; serve with whipped cream or vanilla sauce. Peaches, pears or quinces are made the same way. Finely chopped nuts sprinkled over before the jelly has hardened or freshly grated cocoanut sprinkled over is a great improvement. Half these quantities will be sufficient for a family of 6.

606.**Apples in Jelly, No. 2.**— Pare 1 dozen pippin or greening apples, remove the cores without breaking the fruit and lay the apples into water with the juice of 1 lemon (this will keep them from turning); put the peels and cores of apples into a kettle, cover with water and boil until soft; strain first through cheesecloth and then through a flannel bag till the liquor is clear; return the liquor to kettle and when it boils put in the apples; boil until a straw will easily pierce through them; then remove the apples carefully to a glass dish; measure the apple water and allow 1 pound sugar for 1 pint water; boil the liquid 20 minutes; then add the sugar, boil 3 minutes and let it cool off a little; then pour it over the apples and serve when cold without sauce.

607.**Apples (with Custard).**— Pare, core and cut into quarters 6 large pippin or greening apples; put ½ cup sugar with 171 1 cup water in a wide, low pan over the fire and boil 2 minutes; put in as many apple quarters as will lay in without crowding one another; boil until a straw will pierce through them easily; then take the apples out with a skimmer, lay them in a pudding dish and boil the remaining apples the same way; when the apples are all done and laid in the dish make a custard in the following way:—Beat 5 eggs until very light and add 4 tablespoonfuls sugar, 1 teaspoonful vanilla and 1 quart cold milk; pour this over the apples and bake till the custard is firm; when done remove the dish from oven and serve when cold with a napkin folded around the dish.

608.**Apples (with Currant Jelly).**— Take 6 large greening apples, 1 cup currant jelly and 1 cup sugar; pare, core and cut the apples into quarters and lay them in cold water; put the cores and peels in a saucepan, cover with water and boil till tender; strain them through a jelly bag and return liquid to saucepan; as soon as it boils put in some of the apples (not too many at once, so they do not crowd one another) and boil until a straw will easily pierce through them; then take them out carefully, lay on a dish to cool and boil the remaining apples the same way; when all are boiled again strain the liquid and boil it 20 minutes; then measure it; add to 1 pint liquid 1 pound sugar and stir until sugar is dissolved; then add the currant jelly and stir and boil 2 minutes; then remove from fire; rinse out a mould with cold water, sprinkle with sugar, lay in the apple quarters and pour the liquid when nearly cold over them; set on ice to get firm; in serving turn them onto a dish and lay a circle of whipped cream around the dish or the whites of 3 eggs beaten to a stiff froth and mixed with 1 tablespoonful powdered sugar; put the yolks of the 3 eggs with 2 tablespoonfuls sugar and 1 pint milk in saucepan and stir over the fire till just about to boil; add 1 teaspoonful vanilla and when cold serve with the apples. This makes a pretty dish for supper or dessert.

609.**Steamed Apples.**— Pare and core ½ dozen large tart apples and stick 6 cloves all around into each apple; put them in a pan, put a little sugar into each apple and pour some boiling water 172 in the pan; cover and steam them on top of the stove until a straw will penetrate through them easily; transfer the apples to a dish and set aside to cool; boil the peels and cores in water till soft; then strain through a jelly bag, add the water the apples were boiled in and boil the two together 20 minutes; then add for 1 pint liquid 1 pound sugar and boil 10 minutes more; put 1 teaspoonful currant jelly into each apple and set aside to cool; arrange the apples neatly in a long dish, pour the apple jelly over them and set in a cool place till wanted.

610.**Baked Apples, No. 1.**— Pare some large greening or pippin apples and remove the cores without breaking the fruit; set the apples in a shallow tin pan, fill them with sugar and pour a little water in bottom of pan; set them in a hot oven to bake till done; care should be taken not to have them broken; when done remove them from oven, pile up high in a glass dish and dust with fine sugar.

611.**Baked Apples, No. 2.**— Wash and dry some large tart apples and remove the cores without breaking the fruit; set them in a long pan, add a little water and bake in a hot oven; when done transfer them to a dish, sprinkle over some sugar and serve hot.

612.**Baked Apples and Cocoanut.**— Pare and core 6 large greening apples without breaking them, set in a pan and fill each apple with sugar; boil the peels and cores in water till soft and strain them through a bag; pour the liquor over the apples, cover with another pan and boil on top of stove till they are half done; transfer the apples to another pan, sprinkle them thickly with freshly grated cocoanut mixed with sugar and set in oven to bake till done and to a light brown; in the meantime strain the liquor the apples were boiled in into a saucepan and measure it; allow for 1 pint liquid 1 pound sugar and boil 5 minutes; when the apples are done remove them to a glass dish, pour over the apple syrup and sprinkle a thick layer of fresh cocoanut and sugar over; serve when cold.

173

613. **Pound Sweets (baked).** — Remove the cores from ½ dozen pound sweet apples without breaking them; put them in a long, shallow tin pan, add a little water and bake till tender; remove them to a glass dish, put into each apple 1 teaspoonful apple or currant jelly and dust them over with sugar; serve with or without sweet cream.

614. **Pound Sweets (in Syrup).** — Pare, core and cut into halves ½ dozen large sweet apples and lay them in cold water with a little lemon juice or a little vinegar; put the cores and peels over the fire, cover with water and boil till soft; strain through a coarse bag, return the liquor to saucepan and add the juice of 1 lemon, alittle of the rind and 1 cup sugar; as soon as it boils put in the apples; let them boil until a straw will pierce through them easily; remove the apples to a dish and strain the liquor over them; serve when cold. Afew whole cloves and a piece of cinnamon may be added if the flavor is liked.

615. **Apples Baked with Jelly.** — Pare and core 6 good sized greening or pippin apples and stick cloves all around the top near the opening into each one; put them in a tin pan in a hot oven, add ½ cup water and bake till done; put the peels and cores in a saucepan, cover with water and boil till tender; strain through a jelly bag, return the juice to saucepan and boil 20 minutes; then add for 1 pint juice 1 pound sugar and boil 10 minutes; remove the apples to a glass dish, pour the apple jelly over them and serve cold.

616. **Apple à la Neige.** — Pare, core and cut into quarters 6 large tart apples; put them in saucepan with the peel of ½ lemon, ½ cup water and cover and stew till the apples fall apart; then press them through a coarse sieve or colander (the former is the best), add ½ cup sugar and set aside; when cold beat the whites of 6 eggs to a stiff froth, slowly add the apple sauce and continue the beating for ½ hour; heap it on a glass dish, set lady fingers divided in two all around the dish with bits of currant jelly between and 174 serve with the following sauce: —Stir the yolks with 4 tablespoonfuls sugar and 1 teaspoonful vanilla extract to a cream; add slowly 1 pint cold milk and serve with the above dish. This is a pretty supper dish and a nice dessert. This dish may be made of peaches or cranberries the same way.

617.**Apple Sauce.**— Pare, core and cut into small pieces 12 good sized tart apples, put them into a saucepan with ½ pint water and cover and stew till tender; add 1 cup sugar, press it through a sieve or colander, pour into a glass dish and serve either warm or cold.

618.**Apple Sauce (economical).**— Wash 1 dozen tart apples and cut them into pieces; put them over the fire in a porcelain-lined or agate saucepan, add 1 cup water, cover tightly and stew till tender; when done press them through a sieve or colander (the former is best), sweeten with sugar and serve. Apple sauce made in this way needs only half the apples, and is equally as nice when made right as if the apples were peeled. Apples should never be stewed in rusty tins or iron pots, as they will spoil the appearance of the sauce. Take either a porcelain-lined saucepan, an agate kettle, anew tin kettle or pan or a stone saucepan. Either of these are good for stewing fruit in.

619.**Apples Stewed with Lemons.**— Pare, core and quarter 12 good sized tart apples; put a kettle over the fire with 1 quart water, 1 cup sugar, 1 large lemon cut into thin slices and freed from the pits and boil for a few minutes; then put in the apples and boil until a straw will pierce through them easily; then remove from fire, put the apples into a dish and pour the strained syrup over them; serve either warm or cold.

620.**Apples Stewed Whole with Currants.**— Pare and core 1 dozen medium sized tart apples without breaking them; boil 1 cup sugar with 1 pint water to a syrup, put in the apples and boil till a straw will pierce through them easily; then take out the apples carefully; put ½ cup well washed and dried currants into the syrup 175 and boil 5 minutes; pour it over the apples and serve when cold. Finely cut citron or seedless raisins may be used instead of currants, or use all three together.

621.**Stewed Dried Apples.**— Wash and soak the apples for 1 hour, put them in a saucepan, cover with cold water (or put them on with the water they were soaked in), cover, boil slowly till tender and sweeten them with sugar; serve either hot or cold.

622.**Stewed Evaporated Apples.**— Wash ½ pound evaporated apples in several waters, put them in a saucepan, cover with cold water and boil till tender; add ½ cup sugar and boil for a few minutes; transfer them to a dish and serve either hot or cold; or press the apples when done through a sieve and serve in a glass dish.

623.Peaches Stewed Whole.— Pare 1 quart small peaches; boil 1 cup sugar with 1 cup water for a few minutes, put in the peaches and boil till the fruit shows signs of baking; then remove and when nearly cold pour them into a glass dish and serve cold. Apricots are stewed the same way.

624.Stewed Dried Peaches.— Wash and soak ½ pound dried peaches for several hours in cold water, put them with the water they were soaked in over the fire and boil slowly till tender; add sufficient sugar to sweeten and let them boil for 2 minutes longer; transfer the fruit to a dish and serve cold.

625.Stewed Cherries.— Remove the pits from 2 pounds cherries; boil 1 cup sugar with 1 pint water to a syrup, put in the cherries and boil 3 minutes; pour them into a dish and serve when cold. If the cherries are tasteless add the juice of 1 lemon or use 1 cup water, 1 cup red wine and a stick of cinnamon. If the cherries are small stew them with the pits.

626.Stewed Quinces.— Pare and cut the quinces into quarters and lay them in cold water; put the peels and cores over the fire, cover with water and boil till tender; strain them through a coarse 176 bag made of double cheesecloth; put the liquor over the fire and when it boils add the quinces and boil till they are soft; then add sufficient sugar to sweeten, boil for a few minutes, pour them into a dish and set aside to cool till wanted.

627.Stewed Dried Apricots.— Wash and soak ½ pound dried apricots for 2 hours in cold water, put them with the water they were soaked in over the fire and boil till done; if the water boils away add more; add lastly sufficient sugar to sweeten and serve when cold. Dried cherries are stewed the same way.

628.Baked Pears.— Remove the eyes from 1 dozen nice, ripe pears, put them in a pan with 1 cup water, sprinkle over a little sugar and bake till done; remove them to a glass dish, pour over the syrup from pan, sprinkle with sugar and serve cold.

629.**Stewed Pears.**— Pare 2 dozen stewing pears (if they are large cut them in halves or quarters, if small leave them whole); put a kettle with 1 cup sugar and 2 cups water over the fire and boil a few minutes; put in the pears and stew till done; pour them into a dish and serve cold. If the syrup should be too thin a teaspoonful cornstarch wet with cold water may be added and boiled with them for a few minutes. Alittle claret, cinnamon, lemon juice and rind may also be added if liked.

630.**Stewed Dried Pears.**— Wash and soak 1 pound dried pears for 2 hours; put them over the fire, covered with cold water, add a small stick of cinnamon, alittle lemon juice and peel and boil until nearly done; then add 5 tablespoonfuls sugar and boil till done.

631.**Stewed Dried Prunes.**— Wash 1 pound dried prunes in several waters, put them in a saucepan, cover with cold water, add the juice and rind of 1 lemon and stew till tender; if the water boils away add more; when done add 4 tablespoonfuls sugar, pour them into a glass dish and serve cold. If the liquor of the prunes should be too much or too thin dissolve a little cornstarch in cold 177 water, add it to the prunes and let them boil for a minute; but care must be taken not to get them too thick.

632.**Prunelles.**— Wash ½ pound prunelles in several waters and soak them for 2 hours in cold water; then put them over the fire with the same water they were soaked in to boil slowly till tender; when nearly done add 1 cup sugar and boil till done; pour them into a dish and serve when cold. If not sweet enough add more sugar.

633.**Stewed Cranberries.**— 1 quart cranberries, 2 cups sugar and 1 cup water; wash and pick over the cranberries carefully, put them in a saucepan with 1 cup water, cover and stew till tender; then strain them through a sieve; return the pulp to saucepan and boil for a few minutes; add the sugar and stir and boil just long enough to melt the sugar; rinse out a mould with cold water and sprinkle with granulated sugar; pour the cranberries when nearly cold into the mould and set it in a cool place to get firm.

634.**Stewed Huckleberries.**— Pick over 1 quart huckleberries, put them in a colander and rinse with cold water; put a saucepan with ½ cup sugar and ½ cup water over the fire and add the juice of 1 lemon; when it boils add the huckleberries, cover and stew slowly 10 minutes; add 1 teaspoonful cornstarch wet with cold water and boil for 1 minute; remove the berries from the fire, add ½ cup port wine, pour them into a dish and serve when cold. They are excellent when eaten with German pancake, fried bread or French toast. Or put the huckleberries with ½ cup water, alittle lemon juice (or 1 tablespoonful vinegar) and a small stick of cinnamon in a saucepan over the fire, boil 8 or 10 minutes and sweeten them to taste; break 2 or 3 zwiebacks into small pieces, put them in a dish and pour the huckleberries over them; remove the cinnamon and serve when cold. ½ cup claret added to the huckleberries adds greatly to the flavor.

635.**Stewed Rhubarb.**— Pare and cut 2 bunches rhubarb into small pieces and put over the fire in a porcelain-lined or 178 agate saucepan; add ½ cup water and boil till soft; add 1½ cups sugar or sweeten to taste; boil until the sugar is dissolved, pour into a dish and serve cold.

636.**Stewed Rhubarb (with Eggs).**— Stew the rhubarb the same as in foregoing recipe and take it from the fire; beat 3 eggs to a froth and add 2 tablespoonfuls cold water; then stir in the rhubarb a little at a time; mix well together and serve cold. If stewed rhubarb is too thin add 1 teaspoonful cornstarch wet with cold water and let it boil 1 or 2 minutes.

637.**Compote of Gooseberries.**— Select 1 quart large gooseberries, remove heads and stems and wash and drain them; put them in a kettle, cover with boiling water and boil 5 minutes; pour in a colander to drain; boil 1 cup sugar with ¼ pound water to a syrup and add ¾ cup white wine; put the gooseberries in a porcelain or glass dish and pour the syrup over them; serve when cold.

638.**Compote of Strawberries.**— Press out the juice from 1 pint large, ripe strawberries and mix it with 6 tablespoonfuls sugar; shortly before serving wash and drain 1½ quarts large, ripe strawberries, put them into a glass dish, sprinkle over 3 tablespoonfuls sugar, pour in half of the cold syrup, shake them up, pour over the other half and serve at once.

639.**Compote of Cherries.**— Remove the pits from 2 pounds large cherries; boil 1¼ cups sugar with ¾ cup water to a syrup, put in the cherries and boil 2 minutes; pour them into a dish, cover with paper and set in a cool place; when cool drain off the syrup and reduce it to one-half by boiling it down; then set aside to cool; in serving put the cherries into a glass or fine porcelain dish and pour the cold syrup over them.

640.**Compote of Raspberries.**— Pick over carefully 1½ quarts raspberries, put them in a glass dish and set on ice; shortly before serving sprinkle over 2 tablespoonfuls sugar; press out the juice of 1 pint raspberries, put the liquid with the same quantity of sugar over the fire and boil 10 minutes; let it get cold and pour the 179 syrup just before serving over the raspberries. Currant juice may be used instead of raspberry juice. Acompote may also be prepared with half currants and half raspberries.

641.**Compote of Greengages (or large Egg Plums).**— Select 3 dozen ripe plums, either greengages or the large egg plums, prick them with a needle all around the stem, put them in a kettle with boiling water and let them boil 30 minutes; drain them on a sieve; boil 1½ cups sugar with 1 cup water to a syrup; put the plums in a dish, pour the boiling syrup over, cover with paper and set them in a cool place for 2 hours; then drain off the syrup and reduce to one-half by boiling it down; arrange the plums nicely in a dish and pour the cold syrup over them.

642.**Compote of Plums.**— Choose 3 dozen large blue plums, cut them open on side, remove the pits and pare off the skins; boil 1 cup sugar with 1 cup water, put in the plums and boil a few minutes; pour them into a dish, cover with paper and let them cool; when cold pour the plums onto a sieve and drain off all the liquid; put the syrup over the fire and boil 10 minutes; when cold put the plums into a glass dish and pour the cold syrup over them.

643.**Compote of Oranges.**— Pare and cut 10 large oranges into slices, remove the pits and sprinkle 6 tablespoonfuls sugar over them; let them stand 1 hour; drain off the syrup, put it over the fire, add the juice of 1 lemon and boil slowly 8 minutes; then set aside to cool; just before serving pour the syrup over the oranges and send to table in a glass dish.

644.**Compote of Prunes.**— Wash 1 pound French prunes in several waters, put them in a saucepan, add sufficient red wine to cover, add a small piece of whole cinnamon and the peel of 1 lemon and boil slowly for 2 hours, or until they are soft; when done add 4 tablespoonfuls sugar and as soon as melted remove them from the fire; serve when cold in a glass dish.

645.**Compote of Raisins.**— Remove the pits from 1 pound large raisins, put them in a saucepan with ½ pint water, ½ pint 180 Madeira wine and 1 cup sugar and boil them slowly for 1 hour; serve in a glass dish when cold.

646.**Compote of Dates.**— Cut 1 pound dates open at the side and remove the pits; put the dates with a little Malaga wine, ½ cup water and 4 tablespoonfuls sugar over the fire and boil slowly nearly 1 hour; then serve when cold.

647.**Compote de Marrons.**— Remove the shell from 2 dozen large Italian chestnuts and boil the nuts for a few minutes in water; take them out one at a time and remove the brown skin; boil ½ pound sugar with ½ cup water, put in the chestnuts, pour them together with the syrup into a stone dish, cover and set them in a cool place; next day pour off the syrup, boil it up, add 1 teaspoonful vanilla extract and pour it over the nuts; repeat this once more and serve when cold.

648.**Compote Chaude de Marrons.**— Shortly before serving roast 2 or 3 dozen large Italian chestnuts in the oven, remove the shells and lay the nuts into a hot dish; put 1 gill rum in a small saucepan with 2 tablespoonfuls sugar; let it get hot, pour over the chestnuts, light the rum and bring it burning to table.

649.**Compote of Pineapple.**— Cut a large, ripe pineapple into thin slices, pare them carefully and remove the core in center; boil 1 cup sugar with ½ cup water; lay the pineapple slices into a dish and pour the syrup over them boiling hot; cover and let them stand 2 hours; shortly before serving lay the slices in a glass dish and pour the syrup through a fine sieve over them.

650.**Compote of Quinces.**— Chose ½ dozen large apple quinces, pare and cut them into quarters, remove the cores and lay the quinces in cold water; put the peels and cores in a kettle, cover with water and boil till soft; strain them first through a coarse bag, then through a flannel bag; return the liquor to kettle, add 1 cup sugar, boil for a few minutes, put in the quinces and boil till tender; put them into a dish and strain the syrup over them.

181

651.**Compote of Peaches.**— Cut into halves 20 large, sound peaches, pare them carefully and remove the pits; crack the pits open, take out the kernels, scald them in boiling water and remove the brown skins; place a porcelain-lined or agate kettle with 1 cup sugar and 1 cup water over the fire and boil a few minutes; put in the peaches and kernels and boil from 6-8 minutes; pour them into a deep porcelain dish, cover with paper and set aside; when cold put them in a sieve or colander over the kettle the peaches were boiled in; drain off all the liquid and boil it down to one-half; shortly before serving pile the peaches up high in a glass dish and pour the syrup over them cold.

652.**Compote of Apricots** is made the same as Compote of Peaches.

653.**Compote of Pears.**— Pare and cut 2 dozen medium sized ripe pears into halves, remove the cores and put the pears in cold water with the juice of 1 lemon or 2 tablespoonfuls white vinegar; place a kettle over the fire with 1 pint water, 10 tablespoonfuls sugar, the juice of 1 lemon, half the rind and boil 3 minutes; remove the scum, put in the pears and boil till a straw will pierce through them easily; then pour them in a deep stone bowl, cover with paper and set aside; when cold drain off the liquor and boil it down to one-half; then set aside to cool; shortly before serving pile the pears up high in a glass dish and pour the syrup over them cold. If large pears are taken use a smaller quantity; if small pears are chosen use a larger quantity, leave them whole, cut the stems off half way and pare them.

654.**Compote de Melons.**— Select 1 large muskmelon (not too ripe), cut it in half, remove the seeds with a tablespoon, cut the melon into large pieces, pare off the skin and cut each long piece in two; put the melon pieces in a saucepan, cover with boiling water and boil 5 minutes; take them out with a skimmer and lay the melon pieces on a dish to cool; put a kettle over the fire with 1 pound sugar and 1 pint water and boil a few minutes; put in the melon and boil 20 minutes; pour it into a dish, cover with paper and 182 set in a cool place; when cold drain off the syrup, return it to kettle and boil slowly 10 minutes; then set aside to cool; pile the melon up high in a glass dish and pour the cold syrup over it. Apiece of green ginger root or the juice of 1 lemon may be boiled with the melon.

655.Compote of Watermelon. — Select a nice, ripe watermelon, cut it in half, then into slices, remove the black pits and cut the red part into small pieces; take for 1 quart of such pieces 2 cups sugar, 1 pint water, apiece of green ginger and the juice of 1 lemon; put sugar, ginger and lemon juice over the fire and boil 5 minutes; put in the melon pieces and boil slowly about 20 minutes; pour into a dish and cover with paper; when cold drain off the syrup, return it to kettle and boil 10 to 15 minutes; arrange the melon nicely in a glass dish and pour the cold syrup over it; remove the ginger before sending to table. The white part of the melon may be used for preserving. If the melon is a large one part of it may be used for compote and the other part cut up into slices and served. Athird part may be used as a fruit salad. (See Fruit Salad.)

656.Compote of Apples. — Choose medium sized tart apples, pare and cut them into halves, take out the cores, round the edges and lay them in cold water with lemon juice; boil 1 pound sugar with 1 pint water and the rind and juice of 1 lemon in a wide, low saucepan; put in the apples and let them boil 3 minutes; then turn the apples around, cover the pan and set it on side of stove, where they will stop boiling; let them stand 10 minutes; then thrust a straw through them; if it goes through easily they are done; if not, boil them for a minute longer; remove them from fire and set aside; when cold take the apples out of the syrup and lay them on a sieve; boil the syrup down until it thickens; pile the apples up in a glass dish and pour the syrup over when cold.

657.Compote de Strélity. — Pare, core, cut into quarters and then into fine slices 12 large pippin apples; boil 1 cup sugar with 1 cup Rhine wine, 1 cup water, the juice of 1 lemon and 3 183 ounces finely sliced citron; put in the apples and boil them for a few minutes; then set aside to cool; soak 1 ounce gelatine in 1 cup water 20 minutes; add 1 cup boiling water and stir until dissolved; add it to the apples and boil for 1 minute; then remove from fire; when cold and beginning to thicken rinse out some small cups or forms with cold water, sprinkle each one with granulated sugar, fill them ¾ full of apples and set on ice; boil ¾ cup sugar with ½ cup Rhine wine and a little lemon juice to a syrup and set it in a cool place until wanted; in serving turn the apples out of the cups, put them on a round or oval-shaped dish and pour the syrup over them; or serve the apples with whipped cream.

PIES.

658.**Directions for Making Pies.**— To succeed in making good pastry the following rules should be observed:—Flour should be of the best quality, dry and sifted before using. Butter, unless fresh, should be washed several times in cold water and dried in a napkin. Lard should be sweet, and is best when tried out from leaf lard. If suet is used it should be fresh, chopped fine and freed from all skin. During the process of chopping it should be dredged with flour. Beef dripping should be clarified, and if the dripping has any odor or by-taste a very disagreeable flavor will be imparted to the paste. Strict cleanliness must be observed. All utensils used for pastry making should be clean and kept exclusively for that purpose. Prepare the crust as quickly as possible and do not touch it with your hands any more than necessary. When the crust is ready take a pie plate (agate pie plates are the best) and dust it with flour; do not grease it with butter or lard. Cut off a portion of the crust, roll it out thin, lay it over the plate, press it down lightly with the hand, set the plate in front of you, press with the palms of both hands against the edge of plate and cut the paste which hangs over the edge off with your fingers. The plate is then ready to receive the ingredients of which the pie is to be made. If pumpkin, cocoanut or custard pie is to be made, brush 184 the surface of crust over with beaten egg and sprinkle over 2 tablespoonfuls finely sifted bread or cracker crumbs; then fill in the mixture. This keeps the crust dry and prevents it from being heavy. Pies that are made of stewed or preserved fruit should also be treated the same way. For fine meringue pies the crust should be baked before the mixture is put in. This is done in the following manner:—Line the pie plate with crust and brush the edge over with beaten egg; then roll some pie crust very thin, cut it into strips 1 inch wide and cut one side of the strips into scallops with a knife; wet the edge of crust on the pie plate with beaten egg or water; then lay the strip around the edge of plate so the scallops stand a little above the edge; next lay some thin, buttered brown paper into the plate all over the crust, fill the plate with dry peas and bake it in a medium hot oven till crust is done; then take it from the oven, remove paper and peas, fill in the mixture and bake again till pie is done; draw the pie to front of oven, spread over the meringue and let it remain in oven for a few minutes; then transfer it to a cool place and serve cold. 1 or 2 quarts of dry peas should be kept for this purpose only. They may be put away in a box or glass jar and can then be used several times. If the peas should at any time become rancid from the butter or lard of which the pie crust is made, pour boiling water over them and drain and rub thoroughly with a dry towel; then spread them apart on shallow tins and when dry put away until wanted again. Instead of peas the pie plate may be filled with pieces of stale bread, which can then be used for bread crumbs; but peas are best for this purpose.

659.**Pie Crust (quick and good).** — 2 cups flour, ½ teaspoonful salt, ½ cup ice water and 1 cup lard; sift flour and salt in a bowl, add the lard and chop it fine with a knife in the flour; add the water and mix it with the same knife into a stiff paste; put the paste on a floured board and work it for a few minutes with the knife; take a portion from it and roll it out thin; line a pie plate with it and fill the plate with the ingredients the pie is to be made of; roll out another portion of paste and spread over the top ½ 185 tablespoonful lard; lay this over the pie with the lard side up, press the paste off which hangs over the edge of plate with your hands and place the pie in oven to bake. This crust is excellent, inexpensive and quickly made; sufficient for 2 large pies.

660.**Rich Pie Crust.** — ½ pound flour, ½ teaspoonful salt, ½ pound lard and ½ cup ice water; sift flour and salt into a bowl, add the water and mix it into a paste; put the paste on a floured board and work it thoroughly for 5 minutes, or until it does not stick to the hands; then roll it out into a square about an inch in thickness; also shape the lard into a square, but 1 inch smaller than the paste; lay it in center of paste, fold the paste over and place it for ½ hour on ice; then put it on the board again, dust it under and over with flour, roll it out 3 times as long as wide with a rolling pin, fold over one-third to the center, roll over it once, fold the other end over that, so the paste is three double, roll over it once with the rolling pin, turn the paste around, roll it out again 3 times as long as wide, fold it up the same way as before and set the paste again for ½ hour on ice; repeat the folding and rolling twice more and let it rest each time for ½ hour; when ready to make the pie roll a portion of the paste out very thin, line pie plate as directed with it and fill the plate with the ingredients the pie is to be made of; roll out another portion of the paste, spread the top thickly with lard, lay the paste over the pie with the lard side up and remove the paste which hangs over the edge of plate by pressing against the edge with the palm of your hand; sufficient for 2 large pies.

661.**Fine Pie Crust.**— 1 pound flour, 1 teaspoonful salt, ½ pint ice water, ½ pound lard, 6 ounces butter and the yolk of 1 egg; sift flour and salt in a bowl, add ½ the lard and chop it fine in the flour with a knife; put the yolk in the ice water and beat it with an egg beater till it foams; then add it to the flour and mix it with the same knife into a stiff paste; turn the paste onto a floured board and roll it into a square piece about an inch in thickness; form the remaining lard and the butter also into a square piece, but 1½ 186 inches smaller on all sides; lay it in center of paste, fold it over the lard and butter together, first from right and left, then from and towards you; lay the paste onto a plate and let it rest for ½ hour on ice; then put it on a board, dust under and over with flour and roll it out 3 times as long as wide, rolling always from you; fold over one-third to the center, roll over it once with the rolling pin, fold the other end over that, so the paste is 3 double, roll over it once with a rolling pin, turn the paste around, roll it out again 3 times as long as wide, fold it up the same as above and set the paste again for ½ hour on ice or in a cool place; repeat the rolling out and folding up twice more and let it rest each time ½ hour; after the last rolling let it rest 10 minutes and then use as directed; sufficient for 4 large pies.

662.**Family Pie Crust.**— ½ pound butter, 6 ounces lard, 1 pound flour, 1 cup ice water and ½ teaspoonful salt; sift flour and salt into a bowl, add the lard and chop it up with a broad-bladed knife into the flour until it is very fine; next pour in the ice water and mix it with the same knife into a smooth paste; turn it onto a floured board, dust under and over with flour and roll it out 1 inch in thickness; divide the butter into 3 parts; put 1 part in small bits in regular rows all over the paste; then fold it up, first the right side towards the left, then the left side towards the right, so the paste has 3 thicknesses; turn it around, roll it out again the same way, put over the second part of butter, fold it up and roll out again the same way; put over the third part of butter, fold it up and roll it out twice more; let it rest for 1 hour on ice or in a cool place; then use; while the rolling and folding is going on flour must be sprinkled under and over the paste; the rolling pin must be kept floured, to keep the paste from sticking to it. This paste may be made the day before it is to be used, and if it stands for 2 or 3 days it will not hurt it as long as the paste is kept in a cool place; sufficient for 4 large pies.

663.Fine Paste for Meat Pies, Patties and Baked Apple Dumplings.— Put 1 pound sifted flour on a paste board, make a 187 hollow in center and put in the yolks of 3 eggs and ½ pound butter; work this quickly with your hands into a stiff paste, adding by degrees a little ice water; then roll it out 1 inch in thickness and fold the right and left side to the center, so they meet together; then fold the other two sides over to the center the same way and set it for 1 hour on ice or in a cool place; when that time has expired lay the paste onto a floured board, roll it out 3 times as long as wide, fold one-third over to the center, roll over it once with the rolling pin, fold the other third over that, so the paste has 3 layers, roll out again 3 times as long as wide, fold it up the same way, let it rest for ½ hour and roll and fold it up once more; then use. This paste is excellent for chicken, oyster, pigeon or beefsteak pie; also for baked apple dumplings and fine patties; sufficient for 1 large pie or for 9 apple dumplings.

664.Puff Paste.— 1 pound flour, a pinch of salt, 1 cup ice water and 1 pound butter; sift flour in a bowl, add salt and ice water and mix it into a smooth paste; work it thoroughly on the board with your hands for 5 minutes, cover and set it for 20 minutes on ice; knead the butter well in ice water, to remove the salt, and dry it in a napkin; put the paste onto a floured board and roll it out into a square about 1½ inches in thickness; press the butter flat, also into a square, but smaller than the paste; lay the butter in center of paste and fold the paste first from the right and left side; then from you and towards you over the butter together; turn the paste over with the folded side towards the board; dust under and over a little flour and roll the paste out 3 times as long as wide; fold the lower third over the center and roll over it once with the rolling pin; then fold over that the upper third, so the paste is three double; roll over it once with the rolling pin; turn the paste around, roll it out again 3 times as long as wide, fold it again 3 double, lay it on a plate and set the paste for 1 hour on ice; then roll it out again and fold the same way twice; let it rest for ½ hour on ice; roll and fold it twice more, so the paste has been rolled out and folded up 6 times; after the last rolling let it rest for 20 minutes and then use. The rolling out and folding together must be done 188 with the greatest care, so the layers fit exactly over one another, as the whole success depends upon this. The paste has attained its greatest lightness when rolled and folded together 6 times; if it is rolled out oftener it will loose in lightness; and if it is to be used where lightness is not wanted it must be rolled and folded together from 8 to 10 times. Puff paste is best made in a cool place, and if handy on a marble slab.

665.Short Paste (Mürber Teig—German art).— ½ pound sifted flour, 6 ounces butter, the yolks of 4 eggs, apinch of salt, 1 tablespoonful sugar and the grated rind of ½ lemon; knead the butter in ice water, to remove the salt, and dry it in a napkin; put the flour on a board, make a hollow in center, put in the yolks, lemon, sugar, salt and butter and work it quickly into a smooth dough with your hands; set it on ice for 1 hour before using. Another way:—½ cup butter, the yolks of 3 eggs, 4 tablespoonfuls cream, ½ tablespoonful sugar, apinch of salt and ½ pound flour; mix these ingredients together the same as above. Short paste (or Mürber Teig) is used a great deal in Germany the same as pie crust in America, and is excellent when made right. Avery nice pie is made as follows:—Roll the paste out very thin, cover a pie plate with it (one which is not very deep), cut off what hangs over the edge of plate, spread a thick layer of any kind of fruit marmalade over it, cover with a thin layer of the paste and bake in a quick oven; or bake thin layers of the paste the same as Jelly Cake, and when done lay 2 together with jelly, fruit marmalade or whipped cream between them. Another way to use it is:—Roll the paste out ⅛ of an inch in thickness, cut it into rounds with a cake cutter, brush them over with beaten egg and sprinkle chopped nuts and sugar over them; bake in a medium hot oven and serve with wine.

666.**Neapolitan Paste.**— Scald 5 ounces almonds in boiling water and let them lay for a few minutes; then remove the brown skins and pound the almonds fine in a wedgewood mortar with the yolks of 4 eggs and 1 tablespoonful powdered sugar; sift 1 pound flour on a pastry board, make a hollow in center, put in ¾ pound 189 butter, ½ pound powdered sugar, the almonds, 1 whole egg and 1 yolk and the finely chopped peel of 1 lemon; work this into a stiff paste and set it for 1 hour on ice before using.

667.**Florentinian Paste.**— Sift 1 pound flour on a pastry board, make a hollow in center, put in 1 cup butter, 1 cup sugar, 5 ounces grated chocolate, 2 whole eggs and the yolks of 5 hard boiled eggs rubbed through a sieve; add a little cinnamon and vanilla; knead this into a smooth paste and let it stand for 1 hour in a cool place before using.

668.**Almond Paste.**— Pour boiling water over ½ pound almonds, remove the brown skins, let the almonds lay in cold water for 24 hours and change the water 2 or 3 times; then pound the almonds in a wedgewood mortar with 2 tablespoonfuls water and the juice of ½ lemon; press them through a sieve and mix with ½ pound powdered sugar into a stiff paste; put the paste into a porcelain-lined saucepan and stir over the fire until it loosens itself from bottom of saucepan; remove the paste from the fire and when cold put some powdered sugar on a pastry board, lay the paste on the sugar and work it into a round ball; then set it in a cool place 1 hour before using.

669.**Boiled Paste (Paté à choux).**— Place a saucepan with 1 pint water or milk over the fire and add 1 cup butter, 1½ tablespoonfuls sugar, ¼ teaspoonful salt and the peel of 1 lemon; as soon as it boils sprinkle in slowly, stirring constantly, 1 pint sifted flour; continue stirring until it has formed into a smooth paste and loosens itself from bottom of saucepan; transfer the paste to a dish and let it cool; then mix it by degrees with 8 whole eggs and use for cream cakes, chocolate eclairs and other small cakes.

670.**Nudels.**— Sift 1 cup flour in a bowl and add a pinch of salt, apiece of butter the size of a hazel nut, 1 egg and 1 tablespoonful water; mix this into a stiff paste and work it well on a board so it does not stick to the hands; then divide it into 4 equal parts; roll each part out as thin as paper and let them lay on a 190 board to dry for 10 minutes; then cut them into strips 1 inch wide; lay 4 strips over one another and cut them as fine as possible, like fine straws; when all are cut scatter the nudels all over the board and let them lay till dry; then use or put them away in a box; they will keep for some time. The yolks of 2 eggs may be used instead of 1 whole egg. Nudels are used for puddings or souflées and serve as a dessert; they are also largely used in soup. They should always be put into boiling water, soup or milk and boiled 10 minutes when wanted for use.

671.**Mince Pie.**— 1 pound finely chopped boiled beef, ½ pound finely chopped suet, 1 pound well washed and dried currants, 1 pound stoned raisins, 1 pound finely cut citron, 1 pound sugar, ½ teaspoonful salt, the juice and a little grated rind of 2 oranges, the juice and grated rind of 1 lemon, 1 pint cider, ½ pint brandy, ½ pint sherry wine, 1 teaspoonful ground cloves, 1 teaspoonful cinnamon, 1 teaspoonful mace, 1 grated nutmeg and 3 pounds finely chopped apples; mix all the ingredients well together and use; sufficient for 6 good sized pies. If this mince meat, is to be kept for any length of time omit the apples and fill the mince meat into glass jars; close tightly and keep them in a cool place. It will then keep all winter. When wanted to make pies of take 1 jar at a time and mix the mince meat with an equal portion of chopped apples; line 2 pie plates with rich pie crust, fill them with the mince meat, cover with same crust as directed (see Directions for Pies), cut a small opening in center and one on each side of upper crust and place the pie in a medium hot oven to bake; when done remove it from oven and pour a little good brandy in center, sides and openings and serve warm. Mince pies will keep in a cool place for two weeks, but they should always be put for 10 or 15 minutes in the oven to heat through before serving. For a large quantity of mince meat put 8 pounds beef off the round in a kettle of boiling water, add 1 tablespoonful salt and boil till tender; when done remove the kettle from the fire and set aside to cool; then take out the meat, remove all skin, fat and hard part and chop the meat as fine as possible; then weigh the chopped meat and take for each pound the same ingredients 191 as in above recipe; put it away in well closed jars without the apples.

672.Mock Mince Pie.— 3 finely rolled soda crackers, 1 cup well washed currants, ½ cup stoned raisins, ½ cup finely cut citron, ½ teaspoonful ground cloves, 1 teaspoonful cinnamon, ½ grated nutmeg, ½ teaspoonful salt, 1 tablespoonful butter, ½ cup sugar, ½ cup molasses, ½ cup brandy or wine, the juice of 1 orange and a little grated rind, the juice and grated rind of ½ lemon and ¼ pound dried apples; wash and stew the dried apples till tender; add the ½ cup sugar and sufficient boiling water to make 3 cupfuls stewed apples; set aside to cool; then mix them first with the rolled soda crackers, by degrees with all the other ingredients and use as directed for pies; sufficient for 3 medium sized pies or 2 large ones. Agood plan is to leave a small opening in center of upper crust and when the pies are done pour a little brandy into it.

673.Apple Pie, No. 1.— Line a pie plate with crust as directed (see Directions for Pies); pare, quarter and cut greening or pippin apples into fine slices; fill the plate with apples, sprinkle over some sugar (about 2 tablespoonfuls for a medium sized pie), cover with crust and bake till apples are done and the crust has attained a delicate light brown color. If the flavor is liked a pinch of cinnamon and nutmeg may be added. In the spring of the year the juice of ½ lemon squeezed over the apples of each pie is a great improvement, as the apples have lost a great part of their flavor. Apple pies are best when eaten the same day they are baked. If they stand over till next day they should be put in the oven for about 10 minutes 1 hour before serving. They will then be as good as fresh pies; otherwise the crust is apt to be tough.

674.Apple Pie, No. 2.— Make an apple pie the same as in foregoing recipe; put ½ tablespoonful butter in small bits over the apples, grate over a very little nutmeg and a pinch of cinnamon, add no sugar, leave a small opening in center of upper crust and bake until done; in the meantime boil 1 cup sugar with ½ cup 192 water 5 minutes; when the pie is done put a small funnel in the opening in center of upper crust and pour the syrup carefully through it into the pie; set the pie aside and serve when cold. The pie plate should be deep and large for this pie.

675.Apple Pie, No. 3.— Line a large, deep pie plate with fine pie crust, fill the plate with finely cut tart apples, sprinkle over ½ cup sugar, dust over a little flour and cover with crust; leave a small opening in center of upper crust and bake till done; 10 minutes before the pie is taken from the oven put a small funnel in the opening in the center of upper crust and pour carefully ½ cup sweet hot cider through the funnel into the pie; when done remove the pie from oven and serve when cold. Another way is:—Stew the apple peels and cores in water till tender; then strain them through a coarse bag, return the liquid to saucepan and boil 10 minutes; then add for 1 cup liquid 1 cup sugar, boil 10 minutes longer and pour the hot syrup into the pie in place of cider.

676.**Apple-Citron Pie.**— Line a deep pie plate with rich pie crust, fill it with finely cut tart apples, lay small pieces of butter between the apples, sprinkle over each pie ½ cup sugar and ½ cup finely cut citron, add 2 tablespoonfuls currant or apple jelly, cover with crust and bake till done.

677.**Apple Meringue Pie.**— Press 1 pint stewed apples through a sieve, sweeten to taste and add the juice of ½ lemon, alittle grated nutmeg and the yolks of 4 eggs; line a pie plate with crust, cover with buttered paper, fill the plate with dried peas and bake till crust is a light brown; remove paper and peas, fill in the mixture, return pie to oven and bake till done; in the meantime beat the 4 whites to a stiff froth and add 1 tablespoonful powdered sugar and a little essence of lemon; when pie is done draw it to front of oven, spread over the meringue and let it remain for a few minutes longer in oven; then take it out and serve when cold.

678.**Dried Apple Pie.**— Wash ½ pound dried apples, put them in a saucepan with plenty of cold water, cover and place 193 saucepan over the fire and stew till done; then add 1 cup sugar; pour the apples into a dish and set aside; when cold line 2 pie plates with fine pie crust, brush the surface of crust over with beaten egg and sprinkle over some bread or zwieback crumbs; fill in the stewed apples, cover with crust and bake till done. Dried peaches or apricots can be used the same way.

679.**Tutti Frutti Pie.**— Pare and cut fine 10 large tart apples, put them with 2 tablespoonfuls butter in a saucepan over the fire and add 6 tablespoonfuls sugar, 2 tablespoonfuls finely cut citron, ½ cup seedless raisins, the same quantity of well washed currants and the grated rind of ½ orange or lemon; stir this over the fire till apples are soft and add ½ cup currant or apple jelly; line a pie plate with fine pie crust, fill the plate full with the apples, cover with crust, in which a small opening should be cut in center, and bake till done; boil the peels and cores of apples with a little water till tender; strain them through a jelly bag, return the liquid to saucepan and boil 10 minutes; add to 1 cup liquid 1 cup sugar and boil 5 minutes longer; when pie is done take it from the oven, put a small funnel into the opening in center, pour carefully some of the apple syrup through the funnel into the pie and serve when cold.

680.**Cherry Pie.**— Line a pie plate with crust and remove the pits from 1 quart nice, ripe cherries; fill the fruit into the plate, sprinkle over some sugar and dust over a little flour; cover with top crust, with a small opening in center, and bake in a medium hot oven; in the meantime stew 1 cup cherries in a little water till tender; strain them, return the liquid to saucepan and boil 5 minutes; add to 1 cup liquid 1 cup sugar and continue the boiling for 5 minutes; remove from fire, add a little brandy or wine and pour this syrup, when pie is done, through a funnel into the pie. Another way is:—Do not stone the cherries; after the plate is lined with crust fill it full with cherries, dust over some flour, sprinkle them with sugar and add 3 tablespoonfuls water to each pie; cover with crust and bake till done. Another way is:—Boil ½ cup currant juice with ½ cup sugar for 5 minutes and when the pie (made 194 like the first one) is done pour the currant syrup through a small funnel into the pie. This is an excellent way to give cherries which have not much flavor a nice taste.

681.**Banana Pie.**— 4 large bananas, 1 cup milk, 4 eggs, ½ cup sugar, ½ tablespoonful melted butter and 1 teaspoonful essence of vanilla; remove the skins and press the bananas through a sieve; mix them with the 4 yolks, sugar, milk, melted butter and vanilla; line a deep pie plate with crust, ornament the edge, lay in a piece of buttered brown paper, fill the plate with dry peas or with pieces of stale bread and bake till done; remove it from oven, free the plate from paper and peas, return the plate for a few minutes to oven again, fill in the banana mixture and bake till done; in the meantime beat the whites to a stiff froth and add 1 tablespoonful powdered sugar; draw the pie to front of oven, spread over the meringue, let it remain for a few minutes longer in oven, take it out, set it in a cool place and serve ice cold.

682.**Pineapple Pie.**— Pare 1 ripe pineapple, remove the eyes and hard core in center and chop it fine; line a deep pie plate with fine pie crust, fill it with the finely chopped pineapple, sprinkle over 1 small cup sugar and dust a little flour over; cover the pie with crust and bake a light brown and well done; put the eyes and cores of the pineapple with 6 greening apples cut into pieces in a saucepan, nearly cover them with water and boil till tender; then strain through a bag, return the liquid to saucepan and boil 20 minutes; then add for 1 cup liquid 1 cup sugar, boil 5 minutes and fill the jelly into tumblers. This makes an excellent pineapple jelly.

683.**Pineapple Pie (with Meringue).**— 1 large, ripe pineapple, ½ cup milk, 1 cup sugar, the yolks of 4 eggs, 1 whole egg and ½ tablespoonful butter; pare and grate the pineapple; then mix it with the sugar and other ingredients; melt the butter before adding it; line a deep pie plate with fine pie crust, cover with buttered paper, fill it with dry peas and bake till done; take it from the oven, remove the paper and peas, fill in the pineapple mixture and bake till done; in the meantime make the meringue, as follows:—

195

Beat the 4 whites to a stiff froth and mix it with 1 tablespoonful powdered sugar; when pie is done draw it to front of oven, spread over the meringue and return the pie for a few minutes to oven again till the meringue is a light brown; serve ice cold. This will make 1 large pie, sufficient for a family of 8 persons.

684.**Prune Pie.**— Wash and soak 1 pound prunes for 4 hours in cold water, drain them in a colander, remove the stones, put the prunes in a dish, pour over 1 cup cold water and let them stand over night; next morning line 2 pie plates with crust, put in the prunes with the liquor, sprinkle over some sugar and a little flour, cover with top crust and bake till light brown and well done. Another way is:—Stew the prunes in a little water, remove the stones, sweeten the prunes with sugar, add the juice of ½ lemon and finish as above.

685.**Peach Pie.**— Pare and slice some large, ripe peaches; line a pie plate with crust, fill it with the peaches, sprinkle over some sugar and bake with an upper crust.

686.**Peach Meringue Pie.**— Line a large, deep pie plate with a rich pie crust, brush the surface of crust over with the beaten white of egg and sprinkle over 2 tablespoonfuls finely sifted bread or cracker crumbs; take 1 can preserved peaches, drain off the liquor, put them in the pie plate (with the hollow side up), sprinkle over a little flour, afew spoonfuls sugar, pour over some of the liquor and bake in a medium hot oven till done; in the meantime make the meringue, as follows:—Beat the whites of 5 eggs to a froth, mix them with 2 tablespoonfuls powdered sugar and flavor with a little essence of vanilla; when pie is done draw it to front of oven, spread over the meringue and bake for a few minutes longer; remove it from the oven and set the pie in a cool place; serve cold; sufficient for 10 persons.

687.**Peach Mountain Pie.**— Pare 1 dozen medium sized peaches; line a large deep pie plate with pie crust, fill the plate with the whole peaches, sprinkle over ¾ cup sugar, cover with a thin crust and bake in a medium hot oven.

196

688.**Plum Pie.**— Remove the pits from some ripe plums, sprinkle the fruit thickly with sugar and let them stand for ½ hour; line a pie plate with crust, put in the plums, cover with crust and bake till done.

689.**Cranberry Pie.**— Wash and stew 1 quart cranberries with 1 cup water; when done press them through a colander or coarse sieve, return the cranberries to saucepan, add 2 cups sugar and boil and stir for 5 minutes; then set aside to cool; line a pie plate with fine crust, brush the surface of crust over with the beaten white of egg and sprinkle over 2 tablespoonfuls finely sifted bread crumbs; put in some of the cranberries, about ½ inch thick, and cover with crossbars of crust (lattice-like); bake a light brown and well done; serve cold.

690.Gooseberry Pie.— Top, tail and wash the berries, put them into a pie plate lined with crust, sprinkle plenty of sugar among them, cover with crust and bake till done.

691.Blackberry Pie.— Line a deep pie plate with crust; have some ripe blackberries washed and drained; fill the plate with the berries, sprinkle over some sugar, pour into each pie 1 tablespoonful vinegar, dust over a little flour, cover with crust and bake a light brown and well done.

692.Huckleberry Pie.— Wash and drain some ripe huckleberries; line a pie plate with crust and cover the bottom of crust with 2 tablespoonfuls finely rolled zwieback; next fill the plate with the berries, sprinkle sugar between and over the fruit, add a little more zwieback, cover with crust and bake in a medium hot oven to a light brown and well done; serve cold dusted with sugar.

693.Currant Pie.— Wash and strip some ripe currants and mix them with the same quantity of sugar; line a pie plate with fine pie crust, fill it with the fruit, dust over some flour, cover with top crust, press the edges firmly together and bake till done and to a light brown. Raspberries and currants may be used together for this pie.

197

694.Rhubarb Custard Pie.— Stew 2 cups finely cut rhubarb with 2 cups sugar and ½ cup water; when done strain the rhubarb through a sieve and mix it with 2 well beaten eggs; have a deep pie plate lined with rich pie crust, fill in the mixture, lay fine strips of pie crust across the pie (lattice-like), place the pie in a hot oven and bake till the custard is firm and the crust a light brown.

695.Rhubarb Pie.— Line a pie plate with some rich pie crust; remove the skin from some fresh rhubarb and cut it into fine pieces; take for every cup rhubarb 1 cup sugar, fill it into the plate and dust over a little flour; cover with crust and bake till done, which requires about ½ hour; serve cold dusted with powdered sugar.

696.Rhubarb Meringue Pie.— Place a saucepan with 2 cups finely cut rhubarb, 2 cups sugar and ½ cup water over the fire and stew 20 minutes; when done press the rhubarb through a sieve, add the beaten yolks of 4 eggs and set aside; line a deep pie plate with crust, ornament the edge, cover with buttered paper, fill the plate with dry peas and bake till crust is a light brown; then remove paper and peas, fill the plate with the rhubarb mixture and bake about 15 minutes; beat the 4 whites to a stiff froth and mix them with 1 tablespoonful powdered sugar; draw the pie to front of oven, spread over the meringue, bake for a few minutes longer and serve when cold.

697.Sweet Potato Pie.— 3 medium sized sweet potatoes, 3 eggs, 1 pint milk, ½ cup sugar, 1 tablespoonful butter, 1 teaspoonful ginger, ½ teaspoonful cinnamon, ½ grated nutmeg and ¼ teaspoonful salt; boil the potatoes until done; scrape off the skin and press the potatoes through a sieve or colander; mix them first with the eggs and salt, then add the melted butter, sugar and spice and lastly the milk; line a large, deep pie plate with rich pie crust, fill in the mixture and bake till done; serve when cold.

698.Pumpkin Pie.— Pare and cut a medium sized pumpkin into pieces, remove the pits, put the pumpkin in a kettle, cover with boiling water, add ½ tablespoonful salt and boil till tender; when 198 done put the pumpkin into a colander and drain off all the water; then press it through the colander; measure the strained pumpkin and take for every quart of it 1 pint milk, ½ tablespoonful melted butter, 1½ cups sugar, 3 eggs, 2 teaspoonfuls cornstarch mixed with the milk and 1 teaspoonful ground ginger; mix all the ingredients together; dust some deep pie plates with flour, line them with pie crust and brush the surface of crust all over with beaten egg; roll out some pie crust and cut it into strips 1 inch wide; cut one side of the strips into scallops and lay it around the edge of plate so the scallops stand a little above the edge of plate; brush the strip over with beaten egg and sprinkle 2 tablespoonfuls fine bread or cracker crumbs over the crust (this keeps the pumpkin from sogging into the crust); fill the plate with the pumpkin mixture, grate over the top some nutmeg and bake till done; when the pumpkin is firm to the touch of your finger and a little brown on top the pie is done; remove it from oven, set in a cool place and serve when cold. Amedium sized pumpkin will make 4 medium sized pies. Agood plan if the family is small is to fill some of the boiled pumpkin as soon as done, boiling hot, into glass jars. Close the jars at once and set them in a cool place. When wanted for use open the jar, turn the pumpkin into a colander, drain off all the water, press the pumpkin through a colander and finish the same as above.

699.**Custard Pie.**— Stir 5 eggs with 5 tablespoonfuls sugar to a cream and add 2 tablespoonfuls essence of vanilla or lemon and 1 quart milk; line a large, deep pie plate with crust, brush the surface of crust all over with the beaten white of egg and sprinkle over 2 tablespoonfuls finely sifted bread or cracker crumbs; pour in the custard, grate over the top some nutmeg and bake in a hot oven till custard is firm. Care must be taken to remove the pie as soon as done, otherwise it will curdle. To ascertain when pie is done stick the handle of a teaspoon into center of custard. If no milk is to be seen and the custard is thick the pie is baked. Remove at once and serve ice cold.

700.**Cocoanut Pie.**— To make 2 large pies take 3 pints milk, 6 eggs, 3 cups freshly grated cocoanut, apinch of salt, 1½ cups 199 sugar and 1½ teaspoonfuls essence of vanilla; stir sugar and eggs to a cream and add the milk, salt and flavoring; take a large cocoanut, remove the shell without breaking the cocoanut, pare off the brown skin and grate the cocoanut; add 3 cups of the grated cocoanut to the other ingredients; line 2 large, deep pie plates with crust, brush the surface of crust over with the beaten white of egg and sprinkle over 2 tablespoonfuls fine bread or cracker crumbs; fill the plates with the cocoanut mixture and bake in a hot oven till firm and a light brown on top; when done remove it from oven and serve cold.

701.**Lemon Cocoanut Pie.**— 1 pint milk, 4 large eggs, 1½ cups sugar, the grated rind and juice of 1 large lemon, 2 cups freshly grated cocoanut and ½ tablespoonful butter; put milk and butter in a saucepan to boil; stir the 4 yolks and sugar to a cream and add the grated rind and juice of lemon; when this is well mixed add gradually, stirring constantly, the boiling milk; when cold add the grated cocoanut; in the meantime line a large, deep pie plate with pie crust, ornament the edge with a strip of crust cut into scallops, brush the surface of crust all over with beaten egg and sprinkle over 2 tablespoonfuls fine bread or cracker crumbs; pour in the cocoanut mixture and bake in a hot oven. While the pie is baking prepare the following meringue:—Beat the 4 whites to a stiff froth and add 1 tablespoonful powdered sugar and a little essence of lemon; as soon as the pie is done take it from the oven, spread over the meringue, make it smooth with a broad-bladed knife dipped in water and return the pie to the oven for 2 minutes; then set it in a cool place and serve very cold.

702.Chocolate Cream Pie. — Place a saucepan with 2 tablespoonfuls grated chocolate, 1 pint milk and ½ cup sugar over the fire; add 2 tablespoonfuls cornstarch and stir and boil for a few minutes; remove from the fire and add 1 teaspoonful essence of vanilla, when cold mix it with the yolks of 4 eggs and finish the same as Vanilla Cream Pie.

703.Vanilla Cream Pie. — Line a large, deep pie plate with crust, lay over it a piece of buttered paper, fill the plate with dry 200 peas or pieces of stale bread and bake till crust is a light brown; in the meantime boil 1 cup milk with 1 tablespoonful butter and a pinch of salt; mix 2 tablespoonfuls sifted flour with 1 cup cold milk to a smooth batter and stir it into the boiling milk; continue stirring and boiling for a few minutes; remove it from fire and let it cool a little; stir the yolks of 4 eggs to a cream with 3½ tablespoonfuls sugar; stir this into the above mixture and flavor with 1½ teaspoonfuls essence of vanilla; when the crust is done remove paper and peas, fill in the mixture and bake 10 minutes; in the meantime beat the 4 whites to a stiff froth and mix it with 1 tablespoonful powdered sugar; draw the pie to front of oven and spread over the meringue; set the pie for a few minutes in the oven and serve when cold.

704.Orange Cream Pie. — 1½ cups milk, ¾ cup sugar, 1½ tablespoonfuls cornstarch, ½ tablespoonful butter, 4 eggs, the juice of 3 oranges, the grated rind of 1 and a pinch of salt; put cornstarch, butter, milk and salt in a small saucepan, set it in a vessel of boiling water and stir over the fire till the contents of saucepan thicken; then remove it from fire and set aside; stir the yolks of 4 eggs with the ¾ cup sugar to a cream, add it to the boiled cornstarch and lastly stir in gradually the juice and grated rind of oranges; line a deep pie plate with fine pie crust, lay over it a piece of buttered brown paper, fill the plate with dry peas or pieces of stale bread and bake till crust is a light brown; remove the paper and peas, fill in the cream and bake till' done; in the meantime beat the 4 whites to a stiff froth and add 1 tablespoonful powdered sugar and a little essence of vanilla;

when pie is done draw it to front of oven and spread over the meringue; return it for a few minutes to oven; then set the pie in a cool place and serve ice cold.

705.**Lemon Cream Pie.**— Boil 1 cup milk with ½ tablespoonful butter and a pinch of salt; mix 1½ tablespoonfuls cornstarch with ½ cup cold milk and stir it into the boiling milk; continue stirring and boil for a few minutes; remove it from fire and set aside to cool; stir the yolks of 4 eggs with 1 cup sugar to a 201 cream and add the grated rind of 1 lemon and the juice of 2; stir this into the cold cream; line a deep pie plate with fine pie crust, ornament the edge with a border, cover it with buttered brown paper, fill the plate with dry peas or pieces of stale bread and bake till crust is a light brown; remove the paper and peas, put in the cream mixture and bake about 10 minutes; in the meantime beat the 4 whites to a stiff froth and add 1 tablespoonful powdered sugar and a little essence of lemon; when the pie is done draw it to front of oven and spread over the meringue; return the pie for a few minutes to oven; then set it in a cool place and serve cold.

706.**Fine Lemon Pie (with an Upper Crust).**— The yolks of 3 eggs, 1 whole egg, 1 cup sugar, the juice and grated rind of 1 large lemon and ½ tablespoonful butter; stir the 3 yolks to a cream and add the grated peel and juice of lemon; put the butter in a small saucepan over the fire; as soon as melted add the yolks and stir the whole over the fire to a creamy thickness; then remove from fire; when cold mix it with the sugar and the whole egg; line a pie plate (one which is not very deep) with fine pie crust, brush the surface of crust over with the beaten white of egg, sprinkle over 2 tablespoonfuls fine bread crumbs and put in the lemon mixture; cover with a thin crust and bake in a medium hot oven to a light brown; serve when cold.

707.**Lemon Pie (plain).**— Mix 1 tablespoonful cornstarch with ½ cup cold water, add ½ cup boiling water, ¾ cup sugar and boil for a few minutes; remove from fire, add the juice of 1 lemon, ½ the grated rind, 1 egg and set aside to cool; line a pie plate with crust, put in the mixture, cover with a thin, rich crust and bake a light brown; serve cold dusted with sugar.

708.**Lemon Meringue Pie.**— 5 large eggs, ½ tablespoonful butter, 1 cup sugar and the juice and grated rind of 1 large lemon; beat 4 yolks to a cream and add the grated rind and juice of lemon; put the butter in a small saucepan over the fire; as soon as melted add the beaten yolks and stir over the fire to a creamy thickness; remove it from fire and when cold mix with 1 cup sugar and 1 whole 202 egg; next line a large (not too deep) pie plate with fine pie crust, ornament the edge either with the pastry wheel or lay strips of paste around the edge cut on one side into scallops, brush the surface of crust all over with beaten egg and sprinkle over a little finely sifted bread or cracker crumbs; put in the lemon mixture, put the pie into a medium hot oven and bake till done; in the meantime beat the 4 remaining whites to a stiff froth and add 2 tablespoonfuls sifted powdered sugar and a little grated lemon peel; when the pie is done take it from the oven just long enough to spread over the meringue; return it again to oven for a few minutes and serve when cold.

709.**D'Artois (or Pie of Marmalade).**— Divide ½ pound puff paste into 2 parts; roll one part out into a thin square piece and spread over it, ½ inch thick, apple marmalade ½ inch from the edge; roll out the remaining half into a piece of same size, hold it on the rolling pin and lay over the marmalade; wet the edge of first paste and press the 2 edges together; cut the top paste with a sharp knife into strips, first lengthwise, then crosswise, like lattice work; put in a tin pan and bake in a medium hot oven to a delicate brown; when done dust powdered sugar over and let it remain for a few minutes in the oven to glaze; then remove and serve when cold. Any kind of marmalade or cream may be used.

710.**Allnumettes.**— Roll out 1 pound puff paste 16 inches long and 5 inches wide and spread over a clear icing made as follows:— Beat the white of 1 egg to a froth, add 4 tablespoonfuls powdered sugar, 2 drops lemon juice and beat it for 5 minutes; spread this over the rolled out paste, let it lay for a few minutes, cut it into 8 pieces and bake in a quick oven from 45 to 50 minutes.

711.**Jelly Tarts.**— Roll the puff paste 1 inch in thickness and cut it into rounds with a biscuit cutter; brush a long tin pan over with water, so as just to dampen the pan, and then lay the rounds in the pan with the side which was rolled towards the pan and not too close together; brush the top over with beaten egg, being careful not to let any of the egg run down the sides: then dip a smaller cake cutter into hot water, press it on each round ⅛ of an inch deep 203 to form the cover; the cake cutter must be dipped in the hot water each time for each round; then bake them in a hot oven to a golden color; when the tarts are done take them out, cut the cover loose with a pointed knife and lift it off; hollow the tarts out a little with your finger; when ready to serve fill them with jelly. These tarts may be filled with stewed oysters or clams. They are then called Oyster or Clam Patties.

712.**Tarts.**— Line some small patty pans with rich pie crust or puff paste and fill them with either fruit marmalade of peaches, apricots, cherries or any kind of preserved or stewed fruit; roll the paste which is left out thin and cut it into strips ¼ inch in width; lay them over the tarts like lattice work, brush over with beaten egg and bake in a hot oven.

713.**Peach Tarts.**— Pare and cut some nice, ripe peaches into halves and boil them for 5 minutes in sugar syrup; take them out and set aside to cool; boil the syrup 5 minutes longer; line some patty pans with rich puff paste or pie crust, put into each a piece of buttered paper, fill them with dry peas and bake in a hot oven till nearly done; remove the paper and peas and fill each one with the peaches and a little syrup; return them to the oven again and bake till done; serve when cold.

714.**Apple Tarts.**— Pare and cut into fine slices ½ dozen large tart apples and put them in a small saucepan with 1 tablespoonful butter, 3 tablespoonfuls sugar, ½ cup seedless raisins, currants and finely cut citron; mix, cover and let it simmer over the fire till apples are soft, but not broken; remove them from fire; add 2 tablespoonfuls currant or apple jelly, mix it with the apples and finish the same as Cranberry Tarts.

715.**Rice Tarts.**— Line 12 small forms with rich pie crust or puff paste; put ½ pound rice with cold water over the fire and boil a few minutes; drain the rice in a colander and rinse with cold water; return the rice to saucepan; cover with sweet cream or milk, add 1 tablespoonful butter and boil till tender; remove it 204 from the fire and mix the rice with the yolks of 4 eggs, 3 tablespoonfuls sugar, 1 teaspoonful vanilla extract and lastly the 4 whites beaten to a stiff froth; fill the rice into the small forms and bake in a medium hot oven 20 minutes; when done take them out of the forms, arrange on a long dish with a napkin, dust over with powdered sugar and serve either hot or cold.

716.**Tartelettes of Cherries.**— Roll out 1 pound short paste ¼ inch in thickness and cut out with a tumbler or round cake cutter 20 rounds about 2 inches in diameter; roll out the remaining paste and cut it into long strips ½ inch wide and ⅛ inch thick; lay these strips around the top edge of each round and ornament them with the pastry wheel by pressing small dents in it with the wheel; lay around each a strip of white paper, fasten the two ends with the white of egg (to keep it in its place) and set them in shallow tins; remove the pits from 2 pounds cherries, mix the fruit with sugar and fill them into the tartelettes; then bake in a hot oven; in the meantime crack the pits of cherries, put them over the fire with a little water and boil 5 minutes; then strain; mix in a bowl 1 cup sour cream with 2 rolled zwiebacks, the yolks of 2 eggs, 2 tablespoonfuls sugar and 2 tablespoonfuls of the strained water from the pits; when the tartelettes are nearly done fill 1 spoonful of the above mixture into each one and return them to oven again until the cream is firm; take them from oven, remove the paper and serve when cold dusted with sugar.

717.**Tartelettes** of peaches, plums or apricots are made the same as cherries, but without the cream.

718.**Tartelettes d'apricots.**— Line ½ dozen small patty pans with puff paste or fine pie crust and fill them half full with apricot marmalade; cut 9 apricots in halves, peel them and remove the pits; boil the apricots for 2 minutes in sugar syrup, lay a half apricot into each tartelette and bake in hot oven; in the meantime boil the syrup until thick; crack the pits, take out the kernels, scald them in boiling water, remove the brown skins and divide them into 205 halves; when the tartelettes are done take them out of the form, lay onto each one 2 halves of the kernels and pour a little of the cold syrup into each one; serve when cold.

719.**Tartelettes** of peaches, plums, apples, cherries, currants, raspberries or gooseberries are made the same way as apricots.

720.**Tartelettes of Strawberries.**— Roll out ½ pound puff paste about ⅛ inch in thickness and cut out with a round cake cutter 10 rounds; lay them into 10 small buttered patty pans and press the paste in evenly; let the paste stand a little higher than the pan; mix flour and water to a stiff dough, divide it into 10 equal parts the size of a walnut, roll it out into balls and put into each form a ball; press them in firmly and bake in a hot oven; when done draw them to front of oven, dust over with sugar and return them to oven again so they obtain a glaze; then remove the inside water dough and set the tarts aside to cool; roll out thin the remaining puff paste and cut out small rings; lay them on shallow buttered tins, dust them with sugar and bake in a slow oven; press ½ pint strawberries through a sieve and mix them with 3 tablespoonfuls powdered sugar; wash and drain ½ quart strawberries, put them in a dish, pour the mashed strawberries over the whole fruit and fill them into the tartelettes; lay onto each one 2 rings and serve them on a napkin. Raspberries and currants stewed together may be used instead of strawberries. Stewed cherries, peaches, apricots, plums or any kind of preserved fruit may also be used.

721.**Fleurons of Puff Paste.**— Roll the puff paste out ⅛ inch in thickness, cut it with a cake cutter into shapes of half moons, lay them on tins, brush over with beaten egg and bake in a quick oven. Fleurons are used for garnishing dishes.

722.**Fanchonnettes de creme.**— Line 12 small patty forms with short paste (Mürber Teig) and put into each one a thin layer of fruit marmalade; put 4 whole eggs into a saucepan and beat them to a froth; add 1 pint sweet cream or milk, 3 tablespoonfuls sifted flour, apinch of salt, 1 heaping tablespoonful butter and 4 tablespoonfuls 206 sugar; stir this over the fire till it boils; then add 10 macaroons rolled fine; fill the patty forms ½ full with the boiled cream when cold and bake in a medium hot oven; when done take them out of the form and fill them full with the remaining cream; when cold spread over some marmalade, squirt over the top a little meringue and serve on a napkin.

723.**D'Artois Meringues.**— Prepare ½ pound puff paste, roll out and fold over 10 times; line a large pie plate with it, spread over a thick layer of pineapple marmalade and bake in a medium hot oven; when done draw it to front of oven and spread over a thick meringue; cut the pie into pieces, move them a little apart, lay on each one a few strips of blanched almonds, dust over some sugar and set for 15 minutes into a cool oven to dry.

724.**Condés.**— Chop fine 6 ounces blanched almonds and mix them with 3 tablespoonfuls powdered sugar and the beaten whites of 1 or 2 eggs; roll out some rich puff paste very thin, spread over the almond mixture and dust over some powdered sugar; cut them into finger lengths 2 inches wide and bake in a slow oven.

725.**Small Royal Cakes.**— Prepare a puff paste and roll it out thin about ¼ of an inch in thickness; mix 6 ounces powdered sugar with the beaten whites of 2 eggs; spread this over the rolled out paste, cut it into strips of 1½ inches wide and 3½ inches long, lay them in shallow tin pans and bake in a slow oven to a delicate brown.

726.**Paté à choux.**— Place a saucepan over the fire with 1 pint water or milk, ½ pound butter, 1½ tablespoonfuls sugar, the grated rind of 1 lemon and a pinch of salt; as soon as it boils add slowly ½ pound sifted flour, stirring constantly; stir until it forms into a smooth paste and loosens itself from bottom of saucepan; transfer the paste to a dish and when cooled off a little mix it by degrees with 6 or 8 eggs. This paste should be soft, but must not run apart when dropped on a tin.

207

727.**Chocolate Eclairs.**— Prepare a pâté à choux, put it into a pastry bag or paper funnel and squirt it upon buttered tins in long narrow cakes 4 inches long and 1 inch wide; brush over with beaten egg and bake in a medium hot oven; when done brush them over with boiled chocolate glaze and set for a few minutes in oven again; then set them aside in a cool place; shortly before serving cut each one open on the side and fill with vanilla cream. For cream cakes drop this mixture (by tablespoonfuls) onto buttered tins, not too close together and in the form of round cake; when cold slit them open on one side and fill with vanilla cream.

728.**Canapes.**— Prepare a puff paste and roll it out ½ inch in thickness; cut it out into square pieces of 2 inches wide, cut these again into strips of ½ inch wide, lay them with the cut side in a shallow tin pan, not too close together, and bake in a hot oven; when done draw them to front of oven, dust with sugar and let them remain in oven a few minutes longer to glaze; put two and two together with jelly between; or they may be served single.

729. <u>Cannelous</u>. — Prepare 1 pound puff paste and roll it out 8 times, instead of 6; then take about 18 connelonghölzer (they consist of round pieces of wood about 5 inches long and a finger thick, and can be bought at wholesale confectioneries) and rub each piece of wood over with butter; roll the paste out very thin and cut it into strips of about 1 inch wide and 9 inches long; wind a strip of the paste around each piece of wood, snake-like, brush them over with beaten egg, lay them in shallow tins and bake in a quick oven; when done remove the pieces of wood and when cold fill the cannelous with whipped cream flavored with vanilla and sweetened with sugar.

730. **Vole-au-vent.** — Carefully prepare 1 pound puff paste and roll and fold it 6 times; great care must be taken in doing this, as the whole result depends upon it; after the last rolling let it lay in summer ½ hour on ice, in winter in a cold place; when ready to use roll the paste out 1 inch in thickness, place the dish on which the vol-au-veut is to be served upside down onto the 208 paste and cut off the paste from the dish; turn the paste around and lay it on a tin which has been dampened with water; make with the back of a knife a few dents in it around the edge, brush the top over with beaten egg and make with the point of a knife a slight incision in the paste all round the top about 1½ inches from the edge; this forms the cover; bake in a very hot oven; do not open the oven for 10 minutes; then open and if the vol-au-veut is a light brown cover with paper and bake from ¾ to 1 hour; when done remove the cover, put the vol-au-veut onto the dish it was made to fit and set it for a few minutes in the oven to dry; then fill with either ragouts or fricasseed chickens, birds, rabbits or pigeons, put on the cover and serve; or fill it with fresh or preserved fruit and serve as a dessert.

731. **Vole-au-vent (with Strawberries and Whipped Cream).** — Roll out some carefully made puff paste ¾ inch in thickness and cut it out with a fluted cutter the desired shape, either round or oval; make a slight incision in the paste 1 inch from the edge and bake in

a very hot oven; when nearly done brush it over with white of egg, dust with sugar and put it back in the oven to glaze; when done remove the interior, or soft crumbs, and fill the vol-au-veut shortly before serving with fresh strawberries sweetened with sugar and cover them with whipped cream.

732.**Vole-au-vent (with Currants and Raspberries.)**—Prepare a vol-au-veut the same as in foregoing recipe; strip some large, ripe, cherry currants from their stems, put them in a colander with the same quantity of raspberries, let cold water run over and drain them well; put the fruit into a dish with plenty of sugar, mix them up with 2 silver forks and let it stand in a cool place for several hours; shortly before serving put the fruit into the vol-au-veut, put over the cover, again dust with sugar and serve.

733.**Vole-au-vent (with Peaches and Cream)** is made the same as strawberries. Preserved pineapples, apricots, cherries or plums may be used in the same manner; also oranges peeled and cut into slices, freed from their pits and well sugared. Put into 209 the vol-au-veut and serve either covered with its own cover or whipped cream. Makes an excellent dish for dessert.

734.**Neapolitan Breads.**— Mix a finely chopped orange peel (only the yellow part) with ½ pound neapolitan paste; divide it into small pieces the size of a walnut, roll these lengthwise about ½ finger thick, bread 3 together, brush them over with beaten egg and bake in a medium hot oven; or the paste may be rolled into long, thin rolls, breaded together and then cut into lengths 2½ inches long.

735.**Viennoises.**— Stir ½ cup butter with ½ cup powdered sugar to a cream and add the yolks of 5 eggs, 1 teaspoonful vanilla extract, 1½ cups sifted flour, ¼ cup cornstarch and lastly the beaten whites of 2 eggs; spread this over a sheet of buttered paper ½ inch in thickness, lay the paper in a shallow tin and bake in a slow oven; when done cut the cake into 2 pieces; cover one piece with pineapple or peach marmalade and lay the other piece over it; also cover the top with marmalade and glaze the whole with wine glaze; then cut the cake at once into small, long pieces and set them in the oven again for a few minutes.

736.**Almond Tartelettes.**— Pound ¼ pound blanched almonds with ¼ cup water in a mortar to a paste and press it through a sieve;

mix it with ¼ pound powdered sugar; next add the beaten whites of 4 eggs and the juice of ¼ orange; have ½ pound puff paste prepared, rolled out and folded 10 times; line 8 or 10 small tin patty forms with the paste, fill them ¾ full with the above almond mixture and bake in a medium hot oven; when nearly done draw them to front of oven, dust over some fine sugar and bake till done.

737.**Puites d'amour.**— Prepare 10 ounces puff paste and roll it out ½ inch in thickness; cut out into 20 rounds, about 2 inches in diameter, with a scalloped tin cake cutter; cut a round piece out of the center of each one, so that only a ring remains; roll out the remaining paste the same way, cut out 20 rounds, lay 210 them in a tin pan and brush over with the white of egg; put on the rings, brush the top over with beaten egg and bake in a hot oven; when done dust them over with sugar and let them remain for a few minutes longer in oven to glaze; remove and shortly before serving fill them either with whipped cream sweetened and flavored with vanilla or some preserved fruit. NOTE.—Care must be taken not to get any egg on the outside of tart, as this will prevent the rising.

738.**Neapolitans.**— Take neapolitan paste, roll it out thin, brush over with egg and bake in buttered tins to a light brown color; when done cut the cake into two pieces; spread one piece with fruit marmalade or jelly and lay the other over it; cut the whole into small pieces, pour over a wine or maraschino glaze and set them for a few minutes in the oven.

739.**D'Artois Grilles.**— Prepare ½ pound puff paste, roll it out and fold over 10 times; roll out one-half into a thin, square piece; put this into a square, shallow tin pan, wet the edge with beaten white of egg, spread over a thick layer of apricot or peach marmalade and over this a thick layer of apple marmalade 1 inch from edge; roll out the remaining paste, cut into strips, lay it over the marmalade like lattice work, brush over with beaten egg and bake; when done dust over some sugar, let it remain for a few minutes longer in the oven and cut into pieces when cold. Another way is to line a pie plate with puff paste, spread over a layer of apple marmalade flavored with vanilla and bake; when done spread over a thin layer of apricot marmalade and pour over this a sugar glaze;

make it smooth with a knife and cut into pieces before the glaze becomes hard.

740.**Cream Tarts.**— Line small patty forms with short paste (Mürber Teig), fill them with vanilla cream (see Cream) and bake in a hot oven; when done spread a thin layer of peach marmalade over the cream, pour over the marmalade a little lemon glaze and let them dry for a few minutes in front of oven; mix the beaten white of 1 211 egg with ¾ cup sugar and 3 drops lemon juice; put this into a paper funnel, squirt a wreath over the glaze and put ½ teaspoonful apple jelly in the center.

741.**Fine Pineapple Tarts.**— Line some small patty forms with neapolitan paste and bake in a quick oven to a delicate brown; when done squirt round the edge a rim of meringue, sprinkle finely chopped almonds or pistachio nuts over and let them dry for a few minutes in oven; shortly before serving fill the tarts with finely cut preserved pineapple and pour a little pineapple syrup over them.

742.**Gooseberry Tarts.**— Remove the tops and ends of 1 quart gooseberries, put them in a saucepan, cover with boiling water and let them boil 3 minutes; pour them into a colander, drain off the water and put them into a dish; sprinkle over 1 cup sugar, add a little white wine and let them stand till cold; then finish the same as Cranberry Tarts.

743.**Grape Tarts** are made the same as Cranberry Tarts.

744.**Cranberry Tarts.**— Boil 1 quart cranberries with 1 cup water till they are soft; then press them through a coarse sieve, put the pulp into a saucepan and boil 5 minutes, stirring constantly; then add 1 pound brown sugar and stir until it is dissolved; line 1½ dozen patty pans with puff paste, put into each one a small piece of buttered paper, fill them with dry peas and bake in a hot oven till nearly done; then take them from the oven, remove paper and peas, fill each tart with the stewed cranberries, return them to the oven again and bake till done; serve cold.

745.**Fanchonnettes de pommes.**— Line 12 patty forms with short paste (Mürber Teig), fill them half full with apricot marmalade mixed with apple marmalade and bake in a medium hot oven till done; when done take them out of the form, fill evenly with marmalade and put over the top a meringue; set them in a long, shallow pan and return for a few minutes longer to oven; arrange them on a long dish with napkin and just before serving put a little currant jelly on top of each; serve cold.

212

746.**Darioles à la vanille.**— Line 10 small patty forms with puff paste which has been rolled out 10 times; mix 1 tablespoonful flour and 1 tablespoonful cornstarch with 1½ cups milk or cream and add 1 whole egg, the yolks of 3, 2 tablespoonfuls sugar and 1 teaspoonful vanilla extract; when this is well mixed press it through a sieve and add 6 macaroons broken into small pieces; 20 minutes before serving fill the forms ¾ full with this cream and bake in a hot oven; as soon as the cream is firm draw them to front of oven and dust with sugar; let them remain a minute longer in oven; then remove, take them out of the form and serve at once on a napkin. Grated orange peel may be substituted for vanilla, lemon or almond flavor.

747.**Flan de fruits printaniers.**— Roll out ½ pound short paste (Mürber Teig) ⅛ of an inch thick and lay over it a large pie plate or round tin cover 12 inches in diameter; cut the paste off close to the edge of plate, lift off the plate and put the round piece of paste onto a large buttered tin or thick brown paper; next prepare a warm paste (pâté à choux), put it into a paper funnel and squirt a border 1 inch high on the surface on top of the round close to the edge; then squirt small rills towards the center of round, so that the flau can be cut by these rills and each piece has a border of the paste; then squirt into each compartment 2 rills, so that each piece has 2 compartments; brush the whole over with beaten egg and bake a light brown; take an equal quantity of currants, raspberries and strawberries, sprinkle them thickly with sugar and set in a cool place; then cut some preserved peaches into pieces and add finely cut preserved pineapple and preserved pitted cherries; reduce the liquor of the 3 preserves by boiling it down; then set it aside to cool; when the flau is done remove it from oven and set aside to cool; shortly before serving cut the flau into pieces, lay them on a large, round plate on a napkin (or take a round teatray) and arrange the pieces so that the flau has its original form again; next put into each compartment the fruit; arrange it tastily, pour a little syrup over each one, brush the border over with syrup and serve.

213

748.Flan de Cerises à la creme.— Line a deep jelly cake tin or pie plate with short paste and fill it with pitted red cherries; sprinkle over some sugar and set the flau in a medium hot oven to bake; in the meantime mix together the yolks of 3 eggs, ½ cup sour cream, 6 finely chopped almonds and 2 tablespoonfuls finely rolled zwieback; when the flau is nearly done pour the mixture over and let it bake till done; serve cold dusted with fine sugar.

749.Flan de frangipane.— Line a large, deep pie plate with fine pie crust or short paste, fill it half full of cream frangipane flavored with a little grated rind and juice of orange and bake in a medium hot oven; when done slip the flau onto a large plate; remove the brown crust on top of cream and spread over a layer of marmalade or jelly; fill up the plate with cream, spread over this a thin layer of marmalade and pour over the top a maraschino or wine jelly glaze.

750.Flan de pommes à l'anglaise.— Line a deep pie plate with short paste (Mürber Teig), spread over a thick layer of apricot marmalade, fill up the plate with finely cut tart apples, sprinkle over some sugar and bake in a medium hot oven; when done remove from oven, spread over the top another layer of apricot marmalade and serve cold.

751.Flan de pommes.— Line a deep pie plate or jelly cake tin with short paste (Mürber Teig) and ornament the edge with a pastry wheel; the paste should be about ¼ inch in thickness; pound 6 ounces blanched almonds with the yolks of 3 eggs, put the almonds in a dish and add 4 more yolks, 4 tablespoonfuls sugar, ¼ pound finely rolled macaroons, 1½ tablespoonfuls melted butter and ¼ pound finely cut citron and orange peel mixed; stir this to a cream and add the beaten whites of 6 eggs; fill this mixture into the lined plate and bake 30 minutes in a medium hot oven; as soon as it begins to brown a little on top cover with buttered paper; in the meantime pare, core and cut into quarters 1 dozen pippin or greening apples; place a kettle with ¼ pound sugar, ½ bottle Malaga wine and 1½ cups cherry syrup over the fire; as soon as it 214 boils put in the apples and continue the boiling until they are soft, but not broken; take out the apples carefully and boil the syrup a little longer; when the flau is done take from the oven and when cold lay the apples over it; arrange them nicely, pour the syrup, half warm, over the apples and serve. Flau of Bartlett pears or quinces are made the same way.

752.**Flan d'apricots à la creme meringue.** — Line a large, deep jelly cake tin with short paste (Mürber Teig), ornament the edge with a pastry wheel, fill it half full of cream frangipane flavored with vanilla and bake in a medium hot oven; when done carefully take it out of the pan, lay it on a flat tin and put a layer of preserved apricots over; cover them with cream frangipane, so the flau is evenly full, spread over a thin layer of apricot marmalade and over this a thin meringue; take some meringue in a paper funnel and squirt rills in small squares over the top like lattice work; dust over some fine sugar and place in a slow oven to bake till a light brown; when cold put a little currant jelly into each small square and serve on a napkin.

753.**Flan de peches.** — Line a large, deep jelly cake tin with short paste (Mürber Teig); roll some paste out and cut it into strips of about 1 inch wide and ¼ inch thick; set this around the inside against the rim of tin and ornament it with a pastry wheel; or cut and scallop the strips before putting into the tin so the points of scallops stand a little over the tin.

754.**Flan de peches.** — Roll out ½ pound short paste ⅛ of an inch in thickness, lay a deep jelly cake tin upside down onto the crust and cut the paste off close to the tin with a knife; remove the tin and lay the round piece of paste into the bottom of tin; roll out the remaining paste and cut it into strips a little wider than the rim of the jelly tins, ornament them with a pastry wheel or scallop one side of the strips; fit the strips in neatly inside the rim so the points of scallops stand a little above the edge of tin; fill the tin half full of apricot marmalade mixed with apple marmalade and also spread some of the marmalade on the sides; put the flau in a hot oven and 215 bake till done; in the meantime pare and cut into halves 12 large, ripe peaches and boil them 5 minutes in sugar syrup; remove the peaches and boil the syrup till it begins to thicken; lay the peaches into the syrup again, add 1 teaspoonful vanilla extract and boil them for a few minutes; when the flau is done take it from the oven; remove the brown crust of marmalade and spread over a little fresh marmalade; take the flau out of the jelly tin, slip it onto a large plate and put in the peaches; have the peach kernels blanched and freed from the brown skins, divide in halves and lay them over the peaches; put little bits of currant jelly over the peaches; boil the peach syrup down a little more and pour when cold, just before serving, over the peaches. This may be made of apples, cherries, apricots, plums, pears or any kind of preserved fruit.

STRUDEL, STRAWBERRY SHORT CAKES, BABA, SOLEIL, ETC.

754a.**Strudel Paste.** — Put ½ teaspoonful butter with ½ cup warm milk, alittle salt and the yolk of 1 egg into a bowl and mix it with sufficient sifted flour to make a soft dough; put the dough on to a floured board and work it with the hands for 10 minutes; it should be soft, but not stick to the hands; brush the paste over with a little warm water; rinse out a bowl with boiling water and put it over the paste; let it lay ½ hour; after that time has elapsed cover a kitchen table with a white cloth, dust with flour, put the dough in center of table and pull it out as thin as possible, like paper; when one side is pulled out lay a rolling pin on it while you pull out the other side; then pull out the edges all around, brush it over with melted butter and spread over a filling of stewed preserved or fresh fruit or fruit marmalade; rice, farina, chocolate or creams may also be used; after the filling is put on lift the cloth up at one end and roll the strudel up like a music roll; butter a large, round pan (not too deep) thickly with butter, put one end of the strudel in the center of pan and turn the other end around it so as to give it the shape of a snake curled 216 up; brush the strudel over with melted butter and bake till a light brown and well done in a medium hot oven; when ready slip the strudel into a dish, sprinkle thickly with sugar and serve. It may be eaten hot or cold, with or without sauce. If 2 small strudels are to be made of this quantity the paste should be divided into 2 parts before pulling it apart. It may also be baked in the pan straight.

755.**Rice Strudel.** — Put ½ cup rice with cold water over the fire, boil 5 minutes, drain in a colander and rinse with cold water; return rice to saucepan again with a little salt and 1 cup milk and boil till soft; when done mix it with 1 ounce butter, the yolks of 2 eggs, 2 tablespoonfuls sugar, ¼ teaspoonful cinnamon and 3 ounces seedless raisins; have the strudel paste pulled out on table as directed, brush over with melted butter, spread over the rice, roll it up and finish as directed in foregoing recipe. Serve with the following sauce:—Boil ½ cup sugar with a little water until it begins to turn yellow; remove from the fire and mix it slowly with the whites of 2 eggs, previously beaten to a stiff froth, and add 1 tablespoonful wine and a little vanilla.

756.**Small Rice Strudels.**— Prepare a dough from 1 tablespoonful butter, 3 eggs, alittle salt and sufficient flour to make a stiff dough; roll it out very thin and cut into rounds the size of a tea plate; parboil ½ cup rice in water, drain in a colander and rinse with cold water; return the rice to saucepan again, add 1 pint cream or milk, alittle salt, sugar to taste and boil till tender; stir 1 tablespoonful butter to a cream and add the yolks of 3 eggs, the grated rind of 1 lemon, the cooled off rice and lastly the whites beaten to a stiff froth; spread a layer of rice over each of the rounds of dough, roll them up like omelets, lay them into a buttered pan or dish, pour 1 cup of boiling cream or milk over and bake the strudels in a medium hot oven to a light brown; serve them sprinkled with sugar as a dessert.

757.**Rice Strudel (with Jelly).**— Put little pieces of apple or currant jelly over the rice before rolling it up, otherwise the same as in foregoing recipe.

217

758.**Farina Strudel.**— Boil 1 pint milk with a little salt, sprinkle in slowly ½ cup farina and continue boiling for 10 minutes; when nearly cold stir 1 tablespoonful butter to a cream, add 2 tablespoonfuls sugar, the yolks of 4 eggs, the farina and lastly the whites beaten to a stiff froth; finish the same as Rice Strudel; serve with snow sauce. These quantities will make 2 medium sized strudels, sufficient for 12 persons; serve as dessert.

759.**Farina Strudel (with Cocoanut).**— Prepare the farina the same as in foregoing recipe; spread the farina over the prepared strudel paste and sprinkle over some grated cocoanut; roll up as directed and bake in a buttered pan; serve with cocoanut snow sauce made as follows:—Boil 1 cup sugar with 1 cup water until it begins to turn yellow; remove from fire, add a little boiling water and mix it slowly with the whites of 4 eggs previously beaten to a stiff froth; add 1 teaspoonful vanilla, ½ cup white wine and 1 cup grated coco-anut.

760.**Citron Strudel.**— Stir the yolks of 6 eggs with 4 tablespoonfuls sugar to a cream and add the beaten whites of 3 eggs; have the prepared strudel paste pulled out on a table, brush over first with butter and then spread over the mixture of eggs and sugar; over this spread the remaining 3 whites beaten to a stiff froth; next sprinkle over ¼ pound well washed and dried currants, ¼ pound finely cut citron and roll the strudel up; have ready a buttered pan, put one end of strudel in center of pan and turn the other end around it as directed; brush over with beaten white of egg, sprinkle with sugar and bake till done; serve either with or without wine cream sauce.

761.**Lemon Strudel.**— Mix 6 ounces finely cut citron with ½ pound finely chopped almonds and add 1 cup of sugar, the juice of 3 lemons and the grated rind of 1; let this stand for 1 hour; prepare a strudel paste as directed, pull it out on a table over floured board or tablecloth, brush over with butter, put on the lemon mixture and roll it up; lay the strudel in a buttered pan, twist it around, 218 brush over with water and sprinkle as much sugar over till it lays dry on top; then bake. This strudel is eaten cold and will keep for weeks.

762.**Almond Strudel.**— Stir 2 eggs with ½ cup sugar and the grated rind of ½ lemon to a cream; pull out the strudel paste over a tablecloth, brush over with melted butter, spread over the egg mixture and sprinkle over this ¼ pound grated almonds and ½ cup currants and seedless raisins; roll it up, put into a buttered pan, brush over with melted butter, sprinkle with sugar, pour ½ cup milk or cream into the pan and bake a light brown; serve with wine cream sauce.

763.**Chocolate Almond Strudel.**— Prepare a strudel paste as directed; stir the yolks of 3 eggs with 2 tablespoonfuls sugar to a cream and add the whites beaten to a stiff froth; brush the pulled out strudel paste over with melted butter and then spread over the egg mixture; next sprinkle over 3 ounces finely grated chocolate and 2 ounces finely cut almonds; roll it up and lay into a buttered pan, brush over with beaten egg and bake in a medium hot oven to a light brown; when done sprinkle over some grated chocolate and sugar; serve cold with cream sweetened with sugar.

764.**Chocolate Cream Strudel.**— Boil 2 tablespoonfuls sugar in ½ cup water and add 2 tablespoonfuls grated chocolate which has been melted in oven; stir this over the fire until chocolate is well dissolved and add ½ pint cream and the yolks of 4 eggs; beat this with an egg beater until nearly boiling; remove and set aside to cool; then mix it with the beaten whites; pull out a strudel paste, brush over with melted butter, spread over the chocolate cream and sprinkle over ½ cup finely cut almonds and 2 tablespoonfuls sugar; roll it up, brush over with melted butter and sprinkle with sugar; then bake in a buttered pan. This strudel is best eaten cold. 1 tablespoonful cocoa may be used instead of chocolate.

765.**Apple Strudel.**— Pare, quarter and cut into fine slices ½ dozen greenings, put them in a saucepan with 1 tablespoonful 219 butter, 3 tablespoonfuls sugar and cover and let them stew for a few minutes till apples begin to get soft; then add 2 tablespoonfuls currants, the same quantity of seedless raisins and finely cut citron, alittle grated orange peel and 1 tablespoonful apple or any other kind of fruit jelly; mix all together and set aside; when cold have the strudel paste pulled out over a tablecloth, brush over with melted butter, spread the apples all over it, roll up and finish as directed.

766.**Cherry Strudel.**— Remove the pits from 1 pound ripe cherries, put them over the prepared strudel paste, sprinkle over some sugar, alittle finely rolled zwieback and finish the same as directed; serve dusted with sugar.

767.**Plum Strudel.**— Remove the pits from 1 pound ripe plums, cut them fine, put them over the strudel paste, sprinkle thickly with sugar, dust over a little flour and finish as directed; serve without sauce and dusted thickly with sugar. Peach strudel is made the same way.

768.**Plain Strawberry Shortcake.**— 1 quart flour, ½ teaspoonful salt, ½ cup butter, 2 cups milk and 2 teaspoonfuls baking powder; sift flour, powder and salt into a bowl, add the butter and chop it very fine with a chopper in the flour; then mix it with the milk into a soft dough; divide it into two equal parts and roll them out to the size of a jelly plate; butter a deep jelly tin, put in 1 layer and brush it over with melted butter, put on the other layer and bake in a quick oven; when done remove it from oven, separate the 2 layers with a broad-bladed knife and spread them with butter; mash some fresh strawberries with a silver spoon, cover the bottom layer with the mashed strawberries and sprinkle thickly with powdered sugar; lay on the other layer with the crust side downward, cover with a thick layer of strawberries, sprinkle with sugar and serve with vanilla sauce, sweet cream or the following sauce:—Beat 2 eggs until they foam, add 2 small cups milk, stirring constantly, sweeten to taste and flavor with vanilla extract.

220

769.Strawberry Shortcake, No. 1.— ½ cup butter, 1 cup sugar, 2 eggs, ¾ cup milk and 2 cups prepared flour; stir butter and sugar to a cream and add the eggs 1 at a time; next add the sifted flour and milk alternately; bake in two well buttered jelly tins in a medium hot oven; when done remove and lay them on a napkin which has been dusted with sugar; when cold put a layer onto a plate and cover the cake with fresh strawberries; sprinkle over some sugar, lay over the other layer, cover the top with strawberries, dust with sugar and serve with cold cream or vanilla sauce.

770.Strawberry Shortcake, No. 2.— 1 cup powdered sugar, 3 eggs and 1 cup sifted flour mixed with 1 teaspoonful baking powder and 2 tablespoonfuls water; stir the yolks and sugar to a cream, add water and flour and lastly the beaten whites; bake in 3 layers; when done lay them over one another with strawberries between, sprinkle top well with sugar and serve with cream or vanilla sauce.

771.Peach Shortcake is made the same as Strawberry Shortcake. In place of berries take peaches pared and cut into slices. Or this cake may be made of all kinds of preserved fruit and served either with sweet cream or vanilla sauce.

772.Vienna Bröselcake.— 4 eggs, 5 tablespoonfuls sugar, 3 cups flour, 1 cup warm milk, ½ cup butter, the grated rind of ½ lemon, 1 yeast cake, ¼ teaspoonful salt and a little vanilla; dissolve the yeast in 1 cup milk, add 1 cup sifted flour and mix it into a batter; set it in a warm place to rise; as soon as the sponge is very light stir butter and sugar to a cream and add by degrees the eggs, 1 at a time, stirring a few minutes between each addition; next add salt, lemon or vanilla and lastly the remaining 2 cups sifted flour and the sponge alternately; beat the whole thoroughly with a wooden spoon and set it aside to rise; when light beat it again with a spoon and fill it into a cake mould with tube in center, which should be well buttered and dusted with fine bread, cracker or zwieback crumbs; let it rise again to double its height; in the meantime cut 1 handful almonds into small pieces without removing the 221 brown skin and mix them with 1 handful sugar, a little cinnamon, alittle grated lemon peel and some melted butter; work this with a fork briskly into the dough; when the cake is ready to bake press little dents in it with the handle of a silver spoon, brush over with beaten egg, spread the almond mixture over the top and bake in a medium hot oven 1 hour.

773.**Brioche Cake.**— ¾ pound sifted flour, ½ pound butter, 4 eggs, 2 tablespoonfuls sugar, 1 yeast cake dissolved in ½ cup warm milk and ½ teaspoonful salt; mix ½ cup flour with the salt, yeast and milk into a batter and set it in a warm place to rise until very light; then stir the butter to a cream and add the sugar, the eggs, 1 at a time, stirring a few minutes between each addition; as soon as the batter is light add it gradually to the butter and egg mixture, add the flour and work it with your hands on a floured board into a soft dough; cover and let it rise to double its height; work it thoroughly and let it rise again; when the dough has attained double its size butter a deep, round cake mould and cover the bottom with a round piece of buttered paper; take one-sixth of the dough off and lay it aside; shape the remaining dough into a round loaf and put it into the buttered pan; make a hollow in center; form the small piece of dough into the shape of a pear and put the pointed end into the center of cake; set it to rise to double its size; brush over with the yolk of 1 egg mixed with 1 tablespoonful water and bake in a medium hot oven.

774.**Small Brioche.**— Take the same dough as in foregoing recipe and roll it into small round balls the size of an egg; make a small opening in center of each with a wet finger; put a small ball of dough the size of a hazel nut into each one and set the brioche into a buttered pan; let them rise to double their size; brush over with the yolk of egg diluted with a little cold water and bake in a hot oven; when done brush them over with melted butter.

775.**Baba.** — ¾ pound flour, ½ pound butter, 5 eggs, 2 ounces sugar, the finely chopped peel of ½ lemon, ¼ teaspoonful salt, 1 222 yeast cake dissolved in ¼ cup warm milk, 2 ounces well washed and dried currants, 1½ ounces seedless raisins, 1½ ounces finely cut citron and a little finely cut candied orange peel; mix yeast, milk and ½ cup flour together and set it in a warm place to rise; stir the butter to a cream, add the sugar and next the eggs, 1 at a time, stirring a few minutes between each addition; next add the yeast which was set in a warm place to rise, then the flour and fruit; beat the whole thoroughly with the right hand for 15 minutes; cover with a clean cloth and let it rise to double its size; press it down and let it rise again; then put it into a well buttered form with a tube in center, which should be ¾ full; let it rise till form is full; paste with the white of egg a strip of buttered paper around the top edge of form and bake the cake about 1 hour; when done turn the cake out of form and set it for a few minutes in the oven to dry; in the meantime put ½ cup sugar with ¾ cup Madeira wine over the fire; let it get hot and pour all over the baba; serve either hot or cold on a napkin. Small babas are made of the same dough and baked in small deep forms, otherwise treated the same as above. Instead of Madeira any other kind of wine may be used; also vanilla or pineapple syrup.

776.**Savarin Cake.** — 1 cup lukewarm cream or milk, 2 yeast cake, 3 ounces butter, 4 eggs, ¼ teaspoonful salt, ½ gill Cognac, 2 tablespoonfuls powdered sugar and 10 ounces sifted flour; dissolve the yeast in half the milk and mix it with half the sifted flour into a smooth batter; cover and set it in a warm place to rise, which will take about ½ hour; in the meantime stir butter and sugar to a cream and add the eggs, 1 at a time, stirring a few minutes between each addition; next add alternately the remaining milk, brandy, flour and lastly the batter which has been set in a warm place to rise; beat the whole with the right hand for 15 minutes; then cover with a napkin and let it rise in a warm place; butter a round mould which holds about 1½ to 2 quarts and dust with flour; turn the mould upside down, so the loose flour may fall out; when the dough is very light mix it with 2 ounces finely cut almonds and carefully fill it into the mould; set again in a warm place to rise to double its 223 size; bake in a medium hot oven for ½ hour; to ascertain whether the cake is done or not thrust one end of a knitting needle into center of cake; if the needle comes out clean it is done; if any dough adheres to it the cake must be baked a few minutes longer; as soon as the cake is done turn it onto a round wire grate or sieve and prepare the following glaze:—Place a saucepan with 1 cup sugar and ½ pint cold water over the fire and boil 5 minutes; add 1 glass Jamaica rum, Cognac, sherry wine, kirsch or any other kind of liquor; set the cake with the grate or sieve onto a dish and pour the syrup evenly all over it; pour the syrup which drops from the cake onto the dish back to saucepan again, boil it up and pour over the cake; lift the grate on one side and slide the cake onto a dessert dish; the top may be decorated with preserved cherries or other fruit. The savarin is served as dessert, either hot or cold. Small savarins are baked in small, deep forms and dipped in hot syrup when done.

777.**Soleil.**— 1 pound sifted flour, 10 ounces butter, 2 ounces sugar, the grated rind of 1 lemon, ¼ pound sweet almonds, 2 yeast cake, 8 eggs, apinch of salt and ½ cup cream; dissolve the yeast in a little warm milk; take 1 cup sifted flour and mix it with the yeast into a soft dough; lay this in a bowl with a little warm cream and set it in a warm place to rise to double its size; sift the remaining flour, sugar and salt into a mixing bowl and make a hollow in center; put in by degrees the well beaten eggs and butter and mix and beat the whole with the hands for 15 minutes; pound the almonds with 2 eggs and press them through a sieve; add them with the yeast dough to the above mixture and mix the whole thoroughly together; dust over some flour and set in a warm place to rise to double its size; then fill it into a paper-lined form, which should be ½ filled, and set it to rise till form is nearly full; take a large tin plate and cover the bottom with salt; onto this set the pan with cake and bake in a medium hot oven about 40 minutes; try the cake with a larding or knitting needle; when done remove it from the oven, turn onto a sieve and pour hot vanilla syrup all over it; when cold boil 1 cup sugar with ½ cup water till it forms a thread between 2 fingers; remove from the fire, add 1 tablespoonful raspberry syrup 224 and stir till nearly cold; pour it over the cake and set in oven for 2 minutes; remove instantly and lay a slice of pineapple in center of cake;

cut the pineapple into 4 parts without altering its form and lay long strips of citron around it to imitate the sun.

778.**Compiegne.**— 1 pound sifted flour, 10 ounces butter, 4 whole eggs, the yolks of 4, 2 tablespoonfuls sugar, ½ cup cream, ¼ teaspoonful salt and 2 yeast cake; dissolve the yeast in ½ cup warm milk, add a few spoonfuls of the flour and mix it into a batter; set this in a warm place to rise; in the meantime stir the butter to a cream and add the sugar, salt and eggs by degrees; next the yeast and sifted flour; beat the whole with your hand for 15 minutes; then add the cream, which should be whipped to a stiff froth; butter a large, deep tin form and cover the bottom with a round piece of buttered paper; fill in the cake mixture, which should fill the form ¾ full, paste a piece of paper around top edge of form and let it rise till form is nearly full; bake in a moderate oven; serve either hot or cold. Small compiegnes are baked in small forms and dipped in sugar syrup mixed with wine or liquor.

779.**Damp Nudels.**— 3 cups sifted flour, 1 tablespoonful sugar, ¼ teaspoonful salt, ½ tablespoonful butter, the yolks of 2 eggs, 2 yeast cake and 1 cup warm milk; put flour into a bowl and make a hollow in the center; mix the yeast with ½ cup warm milk, put it into the hollow of flour and mix with a little of the flour to a thick batter; sprinkle a little of the flour over and cover and set in a warm place to rise, which will take about ¾ hour; then add salt, sugar, the yolks and butter, which has been melted in the remaining ½ cup milk; mix the whole into a soft dough and set it in a warm place; when it has risen to double its height cut pieces off with a tablespoon and form them with your hands into round balls; set these onto a floured board covered with a napkin and let them rise to double their size; then brush them over with melted butter; put 1 cup milk, ½ tablespoonful butter and 1 tablespoonful sugar in a pan which is large enough to hold the nudels without touching each other and can be covered tightly; put the pan on top of stove 225 and as soon as the milk boils put in the nudels, cover tightly and let them boil slowly; when the milk has boiled away and the nudels begin to fry in the butter (which can be heard and smelt) draw them to side of stove and let them remain for a few minutes; transfer them to a deep dish and serve with vanilla or snow sauce; the above will make 20 nudels and will take about ½ hour to cook. NOTE.—If the nudels are not all wanted for dessert the remaining ones can be baked in the oven and served for tea as a substitute for biscuits.

FRUIT DUMPLINGS.
(TO BE SERVED AS DESSERT.)

780.**Baked Apple Dumplings, No. 1.**— 1 cup butter, 4 cups sifted flour, ½ teaspoonful salt, the yolks of 3 eggs and ½ cup cold water; put the flour and salt onto a paste board, make a hollow in center and put in the yolks and butter; work this into a stiff paste, adding the water by degrees; roll it out ½ inch in thickness and then fold it up so that 3 layers lay on top of one another; lay the paste onto a plate and let it stand in a cool place or on ice for 1 hour; pare 9 greening or pippin apples and remove the cores without breaking the fruit; next roll out the paste very thin and cut it in 9 squares; put an apple in center of each square and put into each apple 1 teaspoonful apple jelly; bring the corners of the square together, brush the tips with a little white of egg and press them lightly together; set the dumplings into a long tin pan and bake until apples are done; if the oven is too hot cover them with paper; in serving put 1 dumpling for each person onto a tea plate, pour a few spoonfuls cherry wine or lemon sauce around it and put 1 spoonful hard sauce on top of each apple; or they may be served without sauce and dusted with sugar. Baked apple dumplings may be made of pie crust the same way.

781.**Baked Apple Dumplings, No. 2.**— 2 cups sifted flour, ¼ pound lard, 3 ounces butter, ½ teaspoonful salt, the yolk of 1 226 egg and ½ cup ice water; put the flour and ½ the lard and salt into a bowl and chop the lard fine with a knife; beat the yolk and ice water till it foams, add to the flour and mix it with the knife into a smooth, stiff paste; roll it out on a board into a square, put the remaining lard and butter in the center, inclose it with the paste and set for ½ hour on ice or in a cool place; then roll it out 3 times as long as wide, fold together so that 3 layers lay over one another and let it rest for 15 minutes; roll and fold it twice more; then roll it out to ⅛ of an inch in thickness, cut into square pieces and set in the center of each square a peeled and cored tart apple; bring the 4 corners of square together on top, brush the dumplings over with beaten eggs and bake till a light brown and done; serve with hard and wine or lemon sauce. These quantities will make 5 dumplings.

782.**Baked Apple Dumplings, No. 3.**— 1 pint flour, ½ teaspoonful salt, ½ cup cold water and ¾ cup lard; mix flour, salt and water to a paste and work it for a little while on a board; roll it out into a square piece 1 inch in thickness; also shape the lard into a square piece, but smaller than the paste; fold the paste over the lard and set it for a little while in a cool place; then roll it out ½ inch in thickness and fold together so that it is 3 double; roll and fold it twice more; then roll it out ⅛ of an inch in thickness, cut into squares and finish the same as in foregoing recipe.

783.**Baked Apple Dumplings** (with Baking Powder).—2 cups flour, 1 teaspoonful baking powder, ½ cup water or milk, ½ cup butter or lard and ¼ teaspoonful salt; sift flour, salt and baking powder into a bowl, put in the butter and chop it fine; then mix the whole into a soft dough; roll it out ⅛ of an inch in thickness and cut it into squares; put a tart apple, pared and cored, onto each square and put into each apple a small piece of butter and a little sugar; bring the 4 corners of paste together on top of the apple, fasten them with a little white of egg, put in a long, shallow pan and set in a medium hot oven to bake till done; serve with hard or wine sauce; if the oven is too hot cover them with paper. 227 To ascertain when dumplings are done thrust a knitting needle into them. If it penetrates through the apples easily they are done; if not, the baking must be continued until they are done.

784.**Apple Dumplings (bain-Marie).**— 1 pound flour, 1 teaspoonful baking powder, ½ teaspoonful salt, 1 cup cream or milk and ¾ cup butter; sift flour, salt and powder into a bowl, add the milk and mix it into a paste; roll it out on a floured board into a square piece about 1½ inches in thickness; form the butter into a square piece and lay it in the center of paste; fold the paste over it, first from the right and left side, then from you and towards you; put it on a tin plate and set in a cool place for ½ hour; next roll it out 3 times as long as wide, fold it together and lay one end over the middle; then the other end over that, so the paste is 3 double; roll and fold it twice more the same way; then roll it out thin, butter and dust 8 cups with flour and line them with the crust; fill them with finely sliced tart apples and sugar, cover them with the same paste and set the cups in a pan of hot water; put them in the oven to bake; when done turn them onto a dish, dust with sugar and serve with 2 sauces—a wine or nutmeg sauce and a hard sauce. These dumplings can be made with any kind of fruit.

785.**Steamed Apple Dumplings.**— Prepare a fine puff paste, cut it into squares and inclose in each square a nice peeled and cored, juicy apple; lay the dumplings into a steamer, cover tightly, set the steamer over boiling water and let them steam till done, which will take from 1 to 1½ hours; serve with brandy, hard or sherry wine sauce. Instead of puff paste a fine pie crust may be used.

786.**Boiled Apple Dumplings, No. 1.**— 1 cup finely chopped suet, 2 cups prepared flour, 1 egg and ¾ cup water; mix this into a stiff dough, roll it out ⅛ of an inch in thickness and cut into squares; brush each square over with beaten egg and sprinkle over some finely sifted bread crumbs; put in the center of each a nice pared and cored juicy, tart apple; bring the 4 corners of the 228 paste together on top of apple and press them together; brush the dumplings over with beaten egg and dust over them some fine bread crumbs; wring out some small, square pieces of muslin in hot water and dust them with flour; inclose each dumpling in a cloth, pin the 4 corners together, drop them into slightly salted boiling water and boil ¾ hour; serve with cherry wine or nutmeg sauce, or hard or brandy sauce. Dumplings boiled in this way are dry and light. ½ cup of lard may be substituted for the suet.

787.**Boiled Apple Dumplings, No. 2.**— ½ pound lard, 1 pound sifted flour, 1 teaspoonful salt and ½ cup cold water; put flour, lard and salt in a bowl and chop the lard fine with a knife in the flour; add the water and mix it with the knife into a stiff paste; put the paste on a floured board and roll it out about ½ inch in thickness; fold it up and roll it out again about ⅛ of an inch thick; cut into squares and lay in the center of each a pared and cored juicy, tart apple; bring the corners of the squares neatly together and press them lightly; inclose each dumpling in a small square cloth, bring the 4 corners together and fasten with a pin; each cloth should be previously dipped in hot water, wrung out and dusted with flour on the inside before the dumpling is put into it; drop them into boiling water which is slightly salted, boil ½ hour and serve with cherry wine sauce. NOTE.—A good pie crust can also be used for this purpose. Peach, plum and cherry dumplings are made the same way. The above ingredients will make about 8 dumplings. ½ pound finely chopped suet may be used instead of lard.

788.**Lemon Dumplings.**— Mix 2 cups prepared flour with 2 eggs, 1 tablespoonful butter or ½ cup finely chopped suet, 1½ tablespoonfuls sugar, the juice and half the grated rind of 1 lemon and finish the same as Orange Dumplings; serve with cherry wine sauce or prepare a sauce the same as for Orange Dumplings and flavor with lemon instead of oranges.

789.**Orange Dumplings.**— 2 cups prepared flour, 2 eggs, 2 teaspoonfuls butter, 1 tablespoonful sugar and 1 cup water; mix 229 this into a thick batter; pare 3 nice oranges, cut them into small pieces and remove the pits and all the skin, so that there is nothing left but the meat of the oranges; stir this into the batter; have a kettle of boiling water on the stove with a little salt; next drop the mixture with a tablespoon into the boiling water, taking care not to put too much in at once; cover the kettle and boil the dumplings just 10 minutes; remove them with a skimmer to a warm dish and serve at once with the following sauce:—Stir 2 tablespoonfuls butter with 1 cup powdered sugar to a cream and add by degrees 1 egg, 2 tablespoonfuls brandy or wine, the juice of 1 orange and a little grated rind. These dumplings should be removed from the water as soon as done, otherwise they will become heavy.

790.**Fruit Dumplings.**— Stir 1 ounce butter to a cream and add by degrees 3 eggs, 2 tablespoonfuls sugar, 2 tablespoonfuls finely chopped almonds, alittle grated lemon peel, 4 large apples peeled and cut into small slices and ¼ pound bread that has been previously soaked in water and pressed out in a napkin; add cup of bread crumbs, mix all together, drop with a spoon dumplings from this mixture into slightly salted boiling water, boil until done and serve with wine sauce. Cherries, peaches, pears, plums, blackberries or huckleberries may be used instead of apples.

791.**Strawberry Dumplings.**— 1 cup beef suet freed from strings and chopped fine, 3 cups sifted prepared flour, 1 tablespoonful sugar and 2 eggs; mix with ¾ cup cold water to a soft dough; shape into large balls with floured hands, put into a dumpling cloth that is buttered and floured on the inside and fasten the 4 corners of the cloth together with a pin; drop into slightly salted boiling water and boil till done, which will take about ½ hour. The best way to ascertain if dumplings are done is to thrust a knitting needle into them. If no dough clings to it and the needle comes out clean they are done and must be instantly removed from fire, taken from the cloth, laid on a warm dish and served at once; serve with the same sauce as the following recipe.

230

792.**Strawberry Dumplings (another way).**— 2 cups prepared flour, 2 eggs, 1 tablespoonful butter, 1 tablespoonful sugar and 1 cup milk or water; mix this into a stiff batter and stir in lastly 1 cup well washed and drained berries and finish the same as Orange Dumplings; serve with the following sauce:—Stir 2 tablespoonfuls butter with 1 cup powdered sugar to a cream and add the yolks of 2 eggs and 3 tablespoonfuls white wine; when this is well mixed stir 1 cup nice, ripe strawberries into it; put sauce into a glass dish; have the whites beaten to a stiff froth, spread it over the top of sauce and set nice, large strawberries around the edge of dish. These dumplings may be made of all kinds of fruit—peaches, cherries, plums, pears or apricots.

793.**Fine Fruit Dumplings.**— Pare and quarter some peaches, apples or pears and boil them in sugar syrup until done, but in such a way that the pieces stay whole; then take them out with a skimmer and lay on a long dish; in the meantime prepare some dumplings as follows:—Place 1 cup milk with 1 tablespoonful butter over the fire; as soon as it boils stir in 1 cup sifted flour and stir constantly until it has formed into a smooth dough and loosens itself from the bottom of saucepan; after the dough has cooled stir 1 tablespoonful butter to a cream and add by degrees the yolks of 3 eggs, the dough, 1 tablespoonful dry farina and lastly the whites beaten to a stiff froth; form the mixture with a tablespoon into dumplings and drop them into the boiling syrup in which the fruit was boiled; when done take them out with a skimmer and lay in a circle around the fruit; pour the syrup over them and serve. These dumplings may also be made with dried fruit.

794.**Plain Suet Dumplings.**— 1 pint bread crumbs soaked in 1 cup milk, ½ pound suet freed from strings and chopped fine, 4 eggs (whites and yolks beaten separately), 1 tablespoonful sugar, 1 teaspoonful salt and 1 cup prepared flour or flour with ½ teaspoonful Royal baking powder; work into a smooth dough and shape with floured hands into dumplings; boil them inclosed in little, square pieces of muslin or dumpling cloths that have been 231 previously dipped in hot water, wrung out and floured on the inside; boil 40 minutes and serve with strawberry, cherry or wine sauce; or stir any kind of fresh fruit into ½ pint hard sauce and serve it with the dumplings.

795.**Plain Dumplings (with stewed Apples).**— Pare, core and cut into quarters 6 tart apples; boil 1 cup sugar with 2 cups water to a syrup, put in the apples and boil till tender, but do not allow them to break; when done take the apples out with a skimmer and lay them on a dish; mix 1 cup prepared flour with 1 egg, 1 teaspoonful butter and a little water into a thick batter, drop a small portion of the mixture with a teaspoon into the boiling apple syrup and boil 5 minutes; remove them, lay in a circle around the apples and pour the syrup over them. Afew slices of lemon may be boiled with the syrup. This dish can also be made of pears, dried apples or apricots.

796.**Apple Dumplings (with Rice).**— Place ½ pound rice in a saucepan over the fire with cold water, boil 3 minutes and drain in a colander, rinsing with cold water; then put it back on the fire in the same saucepan with 1 pint milk, ½ teaspoonful salt, ½ teaspoonful sugar and a little piece of butter; boil until thick; remove from the fire and mix it with 2 well beaten eggs; dip the dumpling cloths in hot water, wring them out and flour well inside; put 2 spoonfuls of the boiled rice upon each cloth, spread it out smooth and lay in the center of each a peeled and cored apple; fill the opening left by the removal of the core with currant jelly or sugar; draw the 4 corners of the cloth together, bring them to the top of the apple and fasten with pins; drop them into boiling water and boil ½ hour; serve with sweet cream or vanilla, fruit or claret sauces.

CAKES.

797.**Plain Cake.**– 1 cup butter, 1 cup milk, 2 cups sugar, 3 cups prepared flour, 4 eggs and the grated rind of 1 lemon; stir butter and sugar to a light white cream with your right hand; then stir with a silver spoon, add the eggs, 1 at a time, stirring a 232 few minutes between each addition; next add the sifted flour and milk alternately; butter a large, round cake pan and line it with buttered paper; pour in the cake mixture and bake in a medium hot oven for 1 hour; to ascertain if cake is done thrust a knitting needle into center of cake; if it comes out clean the cake is done; if not, the baking must be continued; when done remove the cake from oven and let it stand 10 minutes; then turn it out of pan, remove the paper and set the cake in a cool place or put it when cold in a tin cake box. If plain flour is used take 1½ teaspoonfuls baking powder and sift it with the flour. Measure with a cup which holds half a pint.

798.**Marble Cake.**– Take the same mixture as for Plain Cake and divide it into 3 equal parts; add to one part some red sugar or a little prepared cochineal, to give it a fine pink tint; stir into another part 3 tablespoonfuls grated chocolate and leave the third part plain; butter a large cake pan and line it with buttered paper; fill the pan about ½ inch deep with the plain batter and drop upon this in 3 or 4 places 1 spoonful of the dark and pink batters; pour in more plain batter; then drop in the pink and brown the same way; continue until all is used; the pink may be omitted if the coloring is not handy; bake the same as Plain Cake; when done ice the cake with boiled chocolate glaze.

799.**Nut Cake.**– Prepare a cake batter the same as for Plain Cake, stir in 1 pint shelled walnuts broken into pieces and finish the same as Plain Cake; or stir 3 cups freshly grated cocoanut into the plain cake batter; or stir 1 pint shelled hickory nuts into the plain cake batter; or almonds cut into strips; Brazil nuts may also be used.

800.**Citron Cake.**– Cut ½ pound citron into fine slices and prepare a cake batter the same as for Plain Cake; butter a large, round pan and line it with buttered paper; pour in a layer of cake batter; then a layer of sliced citron; then batter and citron again; continue until all is used; bake in a medium hot oven till done, 233 which will take about 1¼ hours; if the oven should be too hot cover the cake with buttered paper.

801.**Lady Cake.**— 1 cup butter, 2 cups sugar, 1 cup milk, 3 cups prepared flour, the whites of 8 eggs and the grated rind and juice of 1 lemon; stir butter and sugar with your hand to a light white cream and beat the whites to a stiff froth; take a silver spoon and stir the whites, the lemon, sifted flour and milk alternately into the creamed butter and sugar; butter a large mould and line it with buttered paper; pour in the mixture and bake 1 hour. NOTE.—½ pound blanched almonds cut into strips may be stirred into the cake mixture and flavored with vanilla; or 1 pint shelled walnuts broken into pieces or ½ pound finely cut citron can be stirred into the batter and flavored with essence of almonds; ice with clear icing.

802.**Dutchess Cake.**— 1 cup butter, 1 cup milk, 2 cups sugar, 3 cups prepared flour, the yolks of 8 eggs and 2 teaspoonfuls peach extract; stir butter and sugar with the right hand to a light white cream; then stir with a spoon and add the yolks, 2 at a time, stirring a few minutes between each addition; next add the flavoring, the sifted flour and milk alternately; butter a large, round cake pan and line it with buttered paper; pour in the cake mixture and bake in a medium hot oven 1 hour. NOTE.—1 pint shelled walnuts broken into pieces may be stirred into this cake mixture or Brazil nuts may be used; also peanuts broken into pieces.

803.**Fruit Cake.**— Prepare a cake batter the same as for Plain Cake; remove the stones from ½ pound raisins; cut ¼ pound citron into fine slices and mix the raisins and citron with ½ pound well washed and dried currants; dust the fruit with flour, stir it into the cake mixture and finish the same as Plain Cake.

804.**Rich Fruit Cake.**— 2 pounds stoned raisins, 2 pounds seedless raisins, 2 pounds well washed and dried currants, 1 pound finely sliced citron, 1 pound butter, ½ pint good brandy, 1 pint molasses, 1 pound brown sugar, 2 teaspoonfuls grated nutmeg, 2 234 teaspoonfuls ground cinnamon, cloves and mace, 12 eggs and 1 pound flour sifted with 2 teaspoonfuls baking powder; dredge the fruit with flour; stir butter and sugar with the hand to a light white cream; then stir with a wooden spoon and add the eggs, 1 at a time, stirring a few minutes between each addition; next add the molasses, brandy, spice and sifted flour and lastly stir in the fruit; butter 2 large, round cake pans and line them with brown paper; fill in the mixture and bake in a medium hot oven from 3 to 4 hours. Great care must be taken that the oven is just right, as the cake burns very easily.

805.**Orange Layer Cake.**— ½ cup butter, 1 cup sugar, ½ cup milk, the whites of 3 eggs, 1½ cups prepared flour and the grated rind of ½ orange; stir butter and sugar with your right hand to a light white cream and add the grated orange rind; beat the whites to a stiff froth; then add them alternately with the sifted flour and milk to the above mixture; butter 2 large jelly cake tins and line them with buttered tissue paper; put an equal portion of the cake batter into each pan; spread it evenly with a broad-bladed knife dipped in water and bake the cakes in a medium hot oven till a light brown and done, which will take from 15 to 20 minutes; to ascertain when cakes are done thrust a knitting needle into the center of them; if it comes out clean the cakes are done; if any dough adheres to it the baking must be continued; as soon as the cakes are done remove them from the oven; lay a clean cloth or paper on the kitchen table and dust over it some powdered sugar; turn the cakes out of pans upside down onto the cloth and let them lay till cold; in the meantime prepare the filling, as follows: Put in a small saucepan the juice of 1 orange, 1 teaspoonful lemon juice, alittle grated orange peel, 1 teaspoonful butter and the yolks of 3 eggs; set the saucepan in a vessel of boiling water and stir the contents till they thicken; remove from the fire and when cold add ½ cup sugar; lay one layer of cake, bottom side up, on a jelly cake dish and spread over it the orange mixture; lay over the remaining layer, right side up, and dust with powdered sugar; or ice the cake with clear icing; or cover the top of cake with an orange glaze.

235

806.**Lemon Layer Cake.**— 1 cup butter, 1 cup milk, 2 cups sugar, 3 cups prepared flour and the whites of 6 eggs; stir butter and sugar with the right hand to a light white cream, then stir it with a spoon; beat the whites to a stiff froth; add by degrees the sifted flour, the beaten whites and milk alternately to the above mixture; butter 4 good sized jelly tins and line them with buttered paper; then fill in a thin layer of the cake mixture, spread it smooth with a knife and bake in a medium hot oven to a light brown and well done; in the meantime prepare a filling as follows:—Put in a small saucepan the grated rind and juice of 1 lemon, the yolks of 6 eggs, 1 tablespoonful butter and 2 tablespoonfuls water; set the saucepan in a vessel of boiling water and stir till contents thicken; remove from fire when cold, add 1 cup sugar; when cake is done remove it from oven, lay a clean cloth on a table, dust over some powdered sugar and turn the cake out of pan onto the cloth; when cold put 1 layer on a jelly cake dish, bottom side up, and spread over ⅓ of the lemon mixture; put on another layer, upside down, and spread it with the mixture; then treat the third layer the same way; then put on the last layer, right side up, and cover the top with a lemon glaze or dust it with pow-dered sugar. NOTE.—If this cake is not wanted so large it may be divided into 2 cakes, taking 2 layers for each cake; or use half the quantities. Cream or jelly may be used instead of lemon filling.

807.**Chocolate Layer Cake.**— 4 eggs, 1 cup butter, 2 cups sugar, 1 cup milk, 3 cups prepared flour and 1 teaspoonful vanilla; stir but-ter and sugar to a light white cream and add the eggs, 1 at a time, stirring a few minutes between each addition; next add vanilla, the sifted flour and milk alternately; bake in paper lined jelly tins in a quick oven; make 4 layers; in the meantime prepare the following filling:—Beat the whites of 3 eggs to a stiff froth and add 1½ cups powdered sugar, 4 tablespoonfuls Baker's grated chocolate and 1 teaspoonful vanilla; mix all well together and put it between the layers and on top; or put boiled chocolate glaze between the layers and over the top (see Boiled Chocolate Glaze). The 236 top of cake may be ornamented with blanched almonds laid in a circle around the top and some in the center.

808.**Chocolate Cream Cake.**— 1 cup sugar, ½ cup milk, 1½ cups prepared flour, 1 tablespoonful butter, 2 eggs and mix the same as in foregoing recipe; bake in 2 layers in jelly tins; for the cream:—Boil ¾ cup milk and add ½ tablespoonful butter, 2 tablespoonfuls grated chocolate, ½ cup sugar and 1 tablespoonful cornstarch wet with a little cold water; stir and boil for a few minutes; remove from fire and mix with 1 beaten egg and ½ teaspoonful vanilla extract; when cold lay one of the cake layers on a flat dish and spread half the chocolate mixture over it; put on the other layer, spread over the top the remaining chocolate cream and decorate the top with shelled walnuts.

809.**Cocoanut Layer Cake.** — ¾ cup sugar, ½ cup butter, 1½ cups prepared flour, the grated rind and juice of ½ lemon, the whites of 3 eggs and ½ cup milk; stir butter and sugar to a light white cream; beat the whites to a stiff froth and add them by degrees alternately with the sifted flour and milk to the creamed butter and sugar; butter 2 good sized jelly cake tins and line them with buttered paper; put an equal portion in each tin, spread it evenly with a broad-bladed knife dipped in water and bake them in a medium hot oven to a delicate brown color; when done remove them from oven and let them stand for a few minutes; then turn the cakes onto buttered paper to cool; in the meantime grate 1 cocoanut and beat the white of 1 egg to a stiff froth; add ¾ cup powdered sugar and the juice of ½ lemon; lay one cake layer, bottom side up, on a jelly cake dish, spread over half the white icing and sprinkle over a thick layer of the freshly grated cocoanut; put on the remaining layer, right side up, spread over the rest of icing, cover with a thick layer of the cocoanut and sift over some powdered sugar. This cake may be served as a dessert with vanilla sauce.

810.**Lemon Cream Cake.** — ½ cup butter, 1 cup sugar, the whites of 3 eggs, ½ cup milk, 1½ cups prepared flour 237 and the grated rind and juice of ½ lemon; stir butter and sugar to a cream; beat the whites to a stiff froth and add them alternately with the sifted flour and milk to the creamed butter and sugar; add lastly the lemon; butter 2 jelly tins and dust them with cracker dust; put in the mixture, spread it evenly with a knife and bake a light brown; when done put a napkin or clean cloth on the kitchen table and dust with powdered sugar; turn the cakes upside down onto the napkin and let them lay till cold; cream for filling: — Boil ¾ cup milk with 2 tablespoonfuls sugar; dissolve 1 tablespoonful cornstarch in ¼ cup cold milk, stir it into the boiling milk and boil a few minutes; add 1 teaspoonful butter, apinch of salt and remove it from fire; beat up the yolks of 3 eggs with 1 tablespoonful cold milk, stir them into the cornstarch and add 1 teaspoonful essence of lemon; when cold put 1 layer of cake, upside down, onto a plate and spread over the cream; put the other layer over it, right side up, and dust the top with powdered sugar.

811.**Vanilla Cream Cake** is made the same as Lemon Cream Cake, using vanilla flavoring instead of lemon.

812.**Jelly Cake, No. 1.** — 3 eggs, 1 cup powdered sugar, 1 cup sifted flour mixed with 1 teaspoonful baking powder, 2 tablespoonfuls water and the grated rind of ½ lemon; stir sugar and eggs to a cream and add alternately the sifted flour, water and lemon; butter 3 medium sized jelly tins and dust them with finely sifted bread crumbs; put an equal portion of the cake mixture into each tin, spread it evenly and bake in a medium hot oven to a delicate brown; when done remove the cakes from oven.

813.**Jelly Cake, No. 2.** — ¾ cup butter, 2 cups sugar, 1 cup milk, 3 cups flour, 1½ teaspoonfuls baking powder, 4 eggs and 1 teaspoonful essence of lemon; stir butter and sugar with the right hand to a light white cream; then stir with a spoon and add the eggs, 1 at a time, stirring a few minutes between each addition; next add the lemon and then alternately the milk and flour; bake in 3 jelly cake tins in a medium hot oven to a delicate brown color; the tins should be lined with buttered paper; when cold lay the layers over one 238 another with jelly between and dust the top with powdered sugar or ice it with fruit icing.

814.**Jelly Cake, No. 3.** — Stir ¼ pound butter with ½ pound powdered sugar to a light cream and add alternately 1½ cups prepared flour (sifted), the whites of 4 eggs beaten to a stiff froth and 10 drops extract of bitter almonds; butter 2 good sized jelly cake tins and line them with buttered paper; put an equal portion of the cake mixture into each one, spread it evenly with a knife dipped in water and bake to a delicate brown color; when cold arrange in layers with jelly between and sift fine sugar over the top.

815.**Wine Glazed Cake.** — 4 eggs, 1 cup flour, ½ cup sugar, 1 teaspoonful baking powder and the grated rind and juice of ½ lemon; stir eggs and sugar to a cream; sift the flour and baking powder together and add them with the lemon to the above mixture; butter a round cake pan and dust it with fine bread crumbs; pour in the mixture and bake about ½ hour in a moderate oven; for glazing dissolve ½ cup sugar in ½ cup cold water and put it over the fire to boil until the sugar forms a thread between 2 fingers; then add 1 tablespoonful sherry wine, remove it from the fire and stir until a skin forms on top; then slowly pour it over the cake.

816.**Wine Glazed Cream Cake.** — Stir 4 eggs with ½ cup granulated sugar to a cream and add ¾ cup sifted flour in which 1 teaspoonful baking powder has been mixed; bake in a round pan; when done pour over a wine glaze the same as in foregoing recipe and decorate the top with blanched almonds, hazel or walnuts; when cold cut the cake in half with a sharp knife; spread the under half thickly with whipped cream, put the other layer over it and cover the top with whipped cream. NOTE. — This mixture may be baked in a long, shallow pan and before putting it into the oven sprinkle 2 tablespoonfuls granulated sugar over the top. When done cut into squares; or omit the sugar and when done glaze with boiled sugar glaze and cut into squares.

239

817.**Pineapple Cake.** — ½ pound butter, 1 pound powdered sugar, ¾ pound flour, 1 heaping teaspoonful baking powder, ½ pint pineapple syrup, 2 whole eggs, the yolks of 4 and 1 teaspoonful essence of vanilla; wash the butter several times in cold water and dry it in a napkin; put butter and sugar in a mixing bowl and stir with the right hand to a light white cream; then stir with a spoon and add the 2 whole eggs, 1 at a time, stirring a few minutes between each addition; next add the yolks, 1 at a time; sift flour and baking powder together; add the flour and pineapple syrup alternately to the above mixture; butter 3 large, deep jelly cake tins and dust them with flour; put an equal portion of the cake batter into each pan, spread it evenly with a broad-bladed knife dipped in water and bake the cakes in a medium hot oven to a delicate brown; when done remove the cakes from the oven; lay a napkin on a pastry board and dust over some powdered sugar; turn the cake out, upside down, onto the napkin; when cold put one cake, bottom side up, on a cake dish and spread over a layer of pineapple marmalade; put on the last layer, right side up, and cover the top with pineapple glaze made as follows: — Stir ½ pound powdered sugar with 3 or 4 tablespoonfuls pineapple syrup and a few drops of prepared saffron to a stiff sauce; set it for a few minutes over the fire, stirring constantly until lukewarm; then pour it by spoonfuls over the cake and lay some preserved pineapple slices in a circle around the cake; or use candied pineapple.

818.**Wild Rose Cake.** — 1 pound powdered sugar, ¾ pound flour, ½ pound butter, 1 teaspoonful baking powder, the whites of 8 eggs and 1 cup white brandy; sift flour and baking powder together; wash the butter in cold water, to remove the salt, and dry it in a napkin; put butter and sugar in a mixing bowl and stir it with the right hand to a light white cream; beat the whites to a stiff froth and stir them with a spoon in small portions alternately with the flour and brandy into the creamed butter; divide the mixture into 4 equal parts; add to one part a little prepared cochineal, to color it a delicate pink, and flavor with 2 teaspoonfuls rose water; stir into the second part 2 tablespoonfuls cocoa and 1 240 teaspoonful vanilla sugar; add to the third part the yolks of 2 eggs and ½ teaspoonful essence of bitter almonds; leave the fourth part white and flavor it with 1 teaspoonful essence of lemon; take some large, deep jelly cake tins, rub them well inside with butter and dust with flour; put each part of cake mixture into a separate pan and spread the batter smooth with a broad-bladed knife; then bake in a medium hot oven to a delicate brown and well done; lay some clean brown paper or a napkin on a table and dust over some powdered sugar; as soon as one cake is done remove from the oven and let it stand 3 minutes; then turn the pan upside down onto the paper; treat the remaining cakes the same way; as soon as the cakes are cooled off prepare a meringue as follows: — Beat the whites of 5 eggs to a stiff froth and mix them with ½ pound powdered sugar; have ready ½ pound blanched almonds, ½ pound blanched walnuts and ½ pound blanched Brazil nuts; chop the nuts fine; when all is prepared put the cakes together and put the white layer upside down on a jelly dish; spread over the layer ⅓ the meringue and sprinkle over ⅓ the chopped nuts; then put on the dark layer; spread again with meringue and sprinkle with nuts; next put on the yellow layer; spread over the remaining meringue and sprinkle over the nuts; lay the pink layer on top, with the right side up, and cover with the following glaze: — Mix ½ pound powdered sugar with a few spoonfuls red fruit juice or fruit syrup, such as red cherry, raspberry or strawberry syrup; stir the sugar to a thick sauce, set it over the fire and stir constantly until the sugar is lukewarm; then pour it by spoonfuls over the cake; lay blanched almonds and blanched walnuts in a circle around the edge of cake and a few in the center.

819.**Biscuit au beurre.**— ½ pound powdered sugar, ¼ pound sifted flour, ¼ pound cornstarch, ½ pound melted butter, 7 eggs and the grated rind of 1 lemon; beat the 7 whites to a stiff froth and add by degrees the yolks and sugar; set this on a hot stove and beat till it is warm; then remove it and continue the beating until nearly cold; strain the melted butter into a cup and continue the beating with the right hand; hold the cup with the melted 241 butter in the left hand; pour butter into the mixture gradually; next stir in the sifted flour and cornstarch; butter a round cake pan and line it with buttered tissue or waxed paper; pour in the mixture; cover the bottom of a pie plate with salt, set the pan with cake onto this and bake ¾ hour in an oven not too hot; if it browns too much put paper over it.

820.**Fine Sponge Cake.**— 1 pound powdered sugar, 12 eggs and not quite 1 pound (about 2 tablespoonfuls less) of flour; then add the grated rind and juice of 1 lemon; put the flour into a tin pan and set it in front of oven to get warm; put the eggs, sugar and lemon into a deep stone mixing bowl; set the bowl into a large dishpan of hot water in such a way that the bowl is half covered with water; beat the contents of bowl with an egg beater for ¾ hour; then slowly add the flour, continue the beating for a few minutes longer and pour the mixture into a large round pan or 2 medium sized ones; the pan should be previously well buttered and lined with fine brown buttered or waxed paper; bake 1 hour in a slow oven. Sponge cake made according to this recipe is elegant, but care must be taken to follow the instructions exactly. Half these quantities will make a good sized cake. If the oven should be rather hot at the bottom put in a large pie plate with salt, set the pan with cake onto it and bake.

821.**Delicate Sponge Cake.**— 9 eggs, 1½ small cups granulated sugar, 1½ small cups flour (sifted 3 times) and the grated rind and juice of 1 lemon; put the 9 yolks in a bowl and the whites into a deep dish; add to the yolks ½ the sugar and stir them to a cream; beat the whites to a stiff froth and add the remaining sugar, beating constantly; then add slowly, in small portions, the creamed yolks; next the lemon; continue the beating with an egg beater until all is well mixed; then stir in lightly the sifted flour; butter a long, shallow pan and line it with buttered paper; pour in the mixture and bake in a slow oven ¾ hour to a delicate brown; when done carefully remove the cake from oven and let it stand for a few minutes before taking out of the pan. This cake may be iced either with clear icing or wine or fruit glaze.

242

822.**Marguerites.**— ¼ pound flour, ¼ teaspoonful salt, ¼ pound sugar, 1 whole egg, the yolk of 1 egg, 1 heaping teaspoonful anise seed, the grated rind of ½ lemon, ½ cup lukewarm water and 2 tablespoonfuls melted butter; measure after the butter is melted; sift the flour in a bowl, add salt, sugar, grated lemon peel, the eggs and mix the whole with the water into a batter; lastly add the butter and anise seed; put a wafer iron over the fire; when hot brush it over with melted lard; put 1 teaspoonful of the batter in the center of wafer iron, close it and bake the cake a light brown on both sides; as soon as one is done take it from the iron and roll up like a tube; continue to bake the remaining batter the same way; these quantities will make 50 cakes. 2 teaspoonfuls of vanilla sugar may be used instead of anise seed. These cakes may be served either with ice cream, wine or coffee.

823.**Macaroons.**— Scald ½ pound sweet almonds with a few bitter ones, remove the brown skins and spread the nuts out on paper to dry; then pound them in a wedgewood mortar to a paste with the whites of 2 eggs and add 2 teaspoonfuls vanilla sugar or the grated rind of 1 lemon; mix it with ½ pound powdered sugar and 2 tablespoonfuls clear icing made as follows: — Take ½ the white of 1 egg, mix it with 2 tablespoonfuls sifted powdered sugar and stir for a few minutes; then add it to the almonds; mix the whole into a firm paste and form with the hands into small round balls the size of a hickory nut; line some shallow tin pans with brown paper (such as is used for wrapping paper), but do not butter it; set the balls in even rows, one inch apart, on the paper, flatten each one a little with a wet finger and bake them in a medium hot oven to a golden color; when done wet a pastry board with cold water and lay the macaroons on the wet board with the paper side towards the board; after 5 minutes the macaroons may be lifted from the paper, as the dampness loosens them; the above quantities will make 50 macaroons. Great care should be taken to use the exact amount of ingredients here stated.

824.**Cookies.** — 1 cup butter, 2 cups sugar, 2 eggs, the grated rind of 1 lemon and 1 teaspoonful baking powder; stir the butter 243 and sugar to a light white cream and add the eggs 1 at a time; sift the powder with 2 cups flour, add it to the mixture and work the whole with sufficient flour into a stiff dough; roll it out ⅛ inch in thickness, cut with a cake cutter into rounds and bake them a light brown in well buttered shallow tin pans.

825.**Butter Cakes.** — 1 yeast cake 1 quart sifted flour, 1½ pints warm milk or water and 1 teaspoonful salt; dissolve the yeast in a little warm water; sift the flour and salt into a mixing bowl, make a hollow in center, pour in the yeast and water, mix into a batter and let it stand over night (this is called setting a sponge); next morning stir 1 cup sugar with ½ cup lard and ½ cup butter to a cream and add 2 eggs, 1 at a time, stirring a few minutes between each addition; then add the grated rind of 1 lemon and a very little powdered cardamon; mix this thoroughly with the sponge, add sufficient sifted flour to make a soft dough, cover it with a clean cloth and set in a warm place to rise to double its height; then butter 4 long, shallow tin pans (12 inches long, 8 inches wide and 1 inch deep) and dust each one with flour; when the dough has attained the desired lightness divide it into 4 equal parts; roll each part out on the pastry board, put it into the pan, press evenly all over and again set it to rise to top of pan; when ready to bake brush each cake over with melted butter, sprinkle over 2 tablespoonfuls granulated sugar mixed with a little cinnamon and bake in a quick oven to a light brown; as soon as done remove the cakes from the pans and lay them on a long platter, one over the other, with the sugared sides together; when cold serve with coffee.

826.**Butter Cakes (with Baking Powder).** — 2 cups sifted flour, 2 eggs, 1½ teaspoonfuls baking powder, 1 tablespoonful melted butter, 1 cup milk, ½ teaspoonful salt, the grated rind of 1 lemon and ½ cup well washed currants; sift flour, salt and baking powder into a mixing bowl, make a hollow in center, put in the 2 whites and 1 yolk of eggs well beaten, add the lemon, the melted butter and mix it with the milk into a thick batter; lastly stir in the currants; spread the mixture into 2 well greased, shallow tin pans; first brush them over with the remaining yolk, then with 2 244 tablespoonfuls melted butter; mix 3 tablespoonfuls granulated sugar, 2 tablespoonfuls finely cut citron, ½ cup finely chopped almonds together, sprinkle this over the 2 cakes and bake immediately in a quick oven till done and a light brown. Prepared flour may be substituted for baking powder.

827.**Apple Cake.** — Prepare a dough the same as for Butter Cakes and divide it into 6 parts, as the dough for Apple Cake has to be thinner than for Butter Cakes; line 6 shallow tin pans with the dough and set it to rise to double its height; in the meantime pare, core and cut into eighths some large, tart apples and lay them together closely in long rows over the cake; drop 1 tablespoonful melted butter over each cake, sprinkle over some granulated sugar and bake in a hot oven; when done dust with powdered sugar.

828.**Cheese Cake.** — Dissolve ½ yeast cake in ½ pint warm milk and add 1 pound sifted flour, 2 eggs, ½ teaspoonful salt, 2 tablespoonfuls butter and ½ tablespoonful sugar; work this into a soft dough and set it in a warm place to rise to double its height; then roll it out ⅛ of an inch in thickness; butter 2 large cheese or pie plates, cover them with the dough, ornament the edge and let it rise again until light; mix 1 pound fresh pot cheese with 1½ cups thick sweet or sour cream, ¾ cup sugar (or sweeten to taste), 3 eggs and ½ cup currants; when this is well mixed together brush the dough over with melted butter and fill the plate with the cheese mixture; bake in a medium hot oven.

829.**Chrysanthemum Cake.** — ½ pint butter, 1 pint sugar, 1½ pints flour sifted with 1½ teaspoonfuls baking powder, the grated rind of 1 orange, ½ pint milk and the whites of 8 eggs; stir butter and sugar with your right hand to a light white cream; beat the whites to a stiff froth, add them to the creamed butter and mix well together with a spoon; add alternately the flour and milk; then add the grated orange peel and a few drops of cochineal, to color the mixture a delicate pink; butter 3 large jelly tins, dust them with fine bread crumbs, fill in the 245 mixture in equal parts and bake in a medium hot oven; when done remove the cakes from the pans and lay them aside to cool; mix ½ cup powdered sugar with the beaten whites of 2 eggs; spread this icing over the layers and sprinkle them thickly with freshly grated cocoanut; lay the layers over one another, cover the top with pink icing and sprinkle over some cocoanut.

830.**Snowflake Cake.**— 2 cups sugar, 1 cup butter, 1 cup milk, 3 cups prepared flour, the grated rind of 1 lemon and the whites of 6 eggs; stir the butter and sugar to a light white cream; beat the whites to a stiff froth, add them to the creamed butter and add the lemon; then add alternately the milk and sifted flour; bake in 3 layers in large jelly tins; when done remove the cakes from the tins and set aside to cool; beat the whites of 2 eggs to a froth and add ½ cup powdered sugar; spread this over each layer and sprinkle them thickly with freshly grated cocoanut; lay the layers over one another, spread the top layer with the icing, cover it thickly with cocoanut and dust over some powdered sugar.

ROLLS AND BREAD.

831.**Parkerhouse Rolls.**— 2 cups warm milk, 1 yeast cake, 2 cups sifted flour, 1 teaspoonful salt, 2 teaspoonfuls sugar, 2 tablespoonfuls melted butter and 1 egg; dissolve the yeast in a little warm milk; sift the flour into a bowl, add sugar and salt, make a hollow in center and put in the yeast and some of the milk; commence mixing it with the right hand; next add the egg, butter and the remaining milk; set it in a warm place till very light; then work with sufficient sifted flour into a soft dough and let it rise again till very light; then roll it out 1 inch in thickness and cut into rounds with a cake cutter; brush the rounds with melted butter, double them over and set in buttered pans 1 inch apart; let them rise to double their size and bake to a fine golden color; while hot brush them over with melted butter.

246

832.**Bread.**— 2 quarts flour, 1 teaspoonful salt, 2 teaspoonfuls sugar, small tablespoonful lard or butter, 1½ pints lukewarm water, 1 of Fleischmann's yeast cakes. Break the yeast cake into a cup, add 1 teaspoonful sugar and sufficient lukewarm water to fill the cup, set it in a warm place till the yeast rises to the surface. Sift flour, sugar and salt into a bowl, add the lard or butter and rub it fine in the flour; then make a hollow in centre of flour, pour in the yeast and the remaining water, stir it with a spoon into a stiff dough, turn it on to a floured board and work it with the hands and some flour until it does not stick to the hands; return the dough to the bowl, cover and let it stand in a warm place to rise. When the dough has risen to double its size, butter 2 brick-shaped pans or use the crusty bread pans, dust them with flour, divide the dough into 2 equal parts, mould them into loaves on the board, put them into the buttered pans, cover and let stand till the dough is to top of pan, place it in medium hot oven and bake 1 hour. If the bread is to be mixed at night take only ½ yeast cake, otherwise proceed the same as above.

COFFEE.

833.**How to Make Coffee.**— Coffee should always be bought in the bean and ground when wanted. It should never be allowed to boil, as all the fine aroma is thereby lost. The finest, quickest 247 and most economical way to make coffee is by making it in a bag made as follows:—Take a piece of coarse unbleached muslin, about ⅜ yard long and ⅜ yard wide, costing about 5 cents per yard; fold on the bias to a point, sew it together in such a way that the bag has the shape of a funnel and hem it on the top; then place the bag in the coffee pot; let the point hang so that it does not quite reach the bottom; let the top of bag lay over the outside of the coffee pot; then put in 1 cup freshly ground coffee, pour over ½ pint boiling water and let it stand 1 minute; then add 1 quart boiling water and let it stand about 3 minutes on side of stove; have ready the urn in which the coffee is to be served, which should be well rinsed with boiling water, pour in the coffee and serve at once; pour more boiling water over the coffee grounds and let it stand on side of stove for a short while; then serve the same way; the second coffee will be found nearly as good as the first. If the coffee is too strong add more water; if not strong enough add less water, as some like it strong and others do not. Another way is to take 3 heaping tablespoonfuls freshly ground coffee and 6 cups boiling water; grind the coffee as fine as possible; rinse out the coffee pot with boiling water, put in the coffee and pour on enough boiling water to cover it well; let it stand in a warm place for 5 minutes, but do not allow it to boil; then add the remaining boiling water, let it stand for a few minutes and either serve in the same coffee pot in which it was made or strain through a fine sieve into a hot silver or china coffee urn and serve at once. This is also an easy and economical way of making good coffee, but the first-named method is the best.

FRUIT SALADS.

834.**Watermelon Salad.**— Cut a watermelon in two, remove the seeds and break the red part into pieces with a silver fork; put it in layers in a glass dish, sprinkle each layer with sugar and place the dish on ice for 2 hours; when ready to serve pour over ½ pint claret. If objected to the wine may be omitted.

248

835.**How to Serve Watermelon.**— Cut a watermelon in half lengthwise; then cut each half first in two and then into long pieces about 2 inches in thickness; arrange the pieces nicely on an oblong plate and serve. The melon should lay for several hours on ice before being cut, as it is not nice unless cold.

836.**Plum and Peach Salad.**— Choose 1 dozen large egg plums, cut them in two and remove the pits; pare and quarter ½ dozen large, ripe peaches and put them in layers alternately with the plums in a glass dish with 1 cup sugar sprinkled between; place the dish on ice for ½ hour before serving.

837.**Orange and Cocoanut Salad.**— Pare and cut some nice oranges into pieces and remove the seeds; put a layer of the oranges sprinkled with sugar into a glass dish, then a layer of freshly grated cocoanut and next a layer of apple or currant jelly; then oranges again; continue in this way until the dish is filled; place the dish on ice for 1 or 2 hours and serve. If not handy the jelly may be omitted.

838.**Peach and Pear Salad.**— Pare and cut into fine slices 4 large, ripe Bartlett pears; pare and cut into quarters ½ dozen large, ripe peaches; put them with the pears into a glass dish with a layer of whipped cream and sugar between and serve at once.

839.**Peach Salad.**— Pare and cut 1 dozen peaches into quarters, put them into a glass dish, sprinkle sugar between and over them and place the dish on ice for ½ hour before serving.

840.**Banana Salad.**— Cut the fruit into slices, put it into a glass dish with sugar sprinkled between, squeeze over some lemon juice and pour over 1 glass claret; place the dish on ice for 1 or 2 hours before serving.

841.**Raspberry and Currant Salad.**— Remove the stems from 1 pound currants and wash and drain them; also wash and drain 1 quart raspberries and put them into a glass dish with 1½ cups sugar; cover and let them stand for 3 or 4 hours before serving.

249

842.**How to Serve Blackberries.**— Put blackberries into a colander, let cold water run over them and set colander for a few minutes into a dishpan to drain; put the berries into a glass dish with powdered sugar. Huckleberries are prepared the same way.

843.**Banana and Orange Salad.**— Remove the skins from 4 bananas and cut the fruit into slices; pare and cut ½ dozen oranges into small pieces and remove the seeds; put oranges and bananas alternately into a glass dish with sugar sprinkled between, set them on ice for 1 hour and then serve.

844.**Pineapple Salad.**— Pare and quarter 1 ripe pineapple, remove the hard part in center and cut each quarter into fine slices or dice; pare and cut ½ dozen oranges into small slices and remove the pits; put oranges into a glass dish in alternate layers with pineapple with plenty of sugar sprinkled between and place them on ice for 1 or 2 hours before serving. Strawberries with pineapple are prepared the same way.

845.**Cherry Salad.**— Remove the pits from 1 pound cherries, sprinkle with sugar and let them stand for 1 hour; then put them with 1 pint ripe strawberries and a little more sugar into a glass dish, set them on ice for 10 minutes and serve.

SALADS.

846.**Fine Mayonaise, No. 1.**— The yolks of 4 eggs, 8 tablespoonfuls salad oil, 4 tablespoonfuls white vinegar, 1 teaspoonful salt, 2 teaspoonfuls sugar, from 1 to 2 tablespoonfuls French mustard and ½ pint whipped cream or 3 tablespoonfuls condensed milk which is not sweet; put the yolks in a small saucepan and stir them to a cream; then slowly add, stirring constantly, 4 tablespoonfuls oil; when this is well mixed add the 4 spoonfuls vinegar, set the saucepan in a vessel of boiling water and stir over the fire till contents of saucepan begin to thicken; then instantly remove and continue the stirring until 250 cold; then slowly add the remaining 4 spoonfuls oil, stirring constantly; next add the salt and sugar; then the mustard and lastly the cream or condensed milk. This mayonaise is excellent if made exactly according to recipe. These quantities will make a mayonaise sufficient for 10 persons. If the mayonaise is not all wanted at one time fill it into jelly glasses without the cream. It will then keep for some time. The cream can be added when wanted for use.

847.**Mayonaise, No. 2.**— 2 eggs, 4 tablespoonfuls oil, 2 tablespoonfuls vinegar, 1 teaspoonful sugar, ½ teaspoonful salt, 2 tablespoonfuls French mustard and 4 tablespoonfuls whipped cream or 1 tablespoonful condensed milk which is not sweet; put the yolks into a small vessel and stir them to a cream; add by degrees the oil, stirring constantly; then slowly add the vinegar, set the vessel in a saucepan of boiling water and stir and boil till the contents of saucepan begin to thicken; then remove from fire and stir until cold; add the salt, mustard, sugar and vinegar; beat the whites to a stiff froth and slowly add them to the above mixture; stir in the cream or condensed milk just before the salad is to be dressed. These quantities will make a salad sufficient for 8 persons. If it is not to be all used at one time put it into a small glass or stone jar (without the cream or milk) and cover tightly to exclude the air. If kept in a cool place it will keep for some time. When wanted for use add the cream. This mayonaise is both cheap and excellent.

848.**Plain Mayonaise.**— Put some cracked ice into a dishpan and place a bowl in the center of the ice; put the yolks of 4 eggs into a bowl and stir them well with a wooden spoon for 5 minutes; then slowly add ½ bottle best olive oil; add only a few drops at a time and stir constantly; if too much oil is added at one time it will not mix together; if the sauce gets too thick add a little vinegar and lastly a few tablespoonfuls whipped cream, salt and vinegar to taste. Another way is to rub the yolks of 2 hard boiled eggs fine and mix them with 2 raw yolks; otherwise finish the same as foregoing recipe.

251

849.**Sauce Tartare.**— Mix 1 pint mayonaise made as in preceding recipe with 1 tablespoonful French mustard and 1 teaspoonful English mustard mixed, 3 anchovies freed from skins and bones and pressed through a sieve, some finely chopped parsley, small, chopped onion, 4 tablespoonfuls chopped capers, some vinegar and pepper; this sauce is mostly served with cold meat.

850.**Mayonaise without Eggs (economical).**— 1 teaspoonful cornstarch, ½ cup boiling water, 6 tablespoonfuls oil, 3 tablespoonfuls French mustard, 1 teaspoonful salt, 2 teaspoonfuls sugar, 4 tablespoonfuls vinegar and 1 tablespoonful condensed milk which is not sweet; mix the cornstarch with a little cold water, add the ½ cup boiling water and stir and boil for a few minutes; then set aside to cool; put the mustard into a bowl and gradually add the oil, stirring constantly; next add the sugar, salt and vinegar; then the cornstarch and lastly the milk.

851.**Salad Dressing without Oil, No. 1.**— 2 ounces butter, 2 eggs, 2 tablespoonfuls vinegar, 1 teaspoonful sugar, ½ teaspoonful salt and 2 tablespoonfuls French mustard; melt the butter in a cup by setting it into hot water for a few minutes and then set it aside to cool; stir the yolks in a small vessel with salt and sugar to a cream and add the melted butter, alittle at a time, stirring constantly; next slowly add the vinegar, set the vessel in a saucepan of hot water and stir until the contents begin to thicken; then remove, stir until cold and slowly add the other ingredients; beat the whites to a stiff froth and mix them with the sauce; if handy add a few tablespoonfuls cream; if English mustard is used take 1 teaspoonful mixed with cold water.

852.**Salad Dressing without Oil, No. 2.**— 1 teaspoonful flour, ½ cup boiling water, 1 tablespoonful butter, 2 tablespoonfuls vinegar,

1 teaspoonful sugar, ½ teaspoonful salt and 2 tablespoonfuls French mustard; mix the flour with a little cold water, add the boiling water, boil 2 minutes and add the butter in small pieces; remove from fire and mix by degrees with the vinegar, then the mustard and the other ingredients; to this sauce the yolks of 2 eggs may be 252 added and also the 2 whites beaten to a stiff froth; or the yolks of 2 hard boiled eggs rubbed through a sieve and the whites chopped fine and sprinkled over the salad.

853.**How to Prepare Lettuce for Salad and for Garnishing.**— Cut off the stalks from 3 or 4 heads of lettuce, pick off all the decayed and withered leaves, break the tender green leaves apart one by one and remove the thick veins; put the lettuce into cold water, rinse well and let it lay in ice water for ½ hour or longer; shortly before serving drain the lettuce in a colander; then put it in a napkin, shake out well and use as directed.

854.**Lettuce Salad (plain).**— Prepare the lettuce as in foregoing recipe, lay it into a salad dish and pour over 2 or 3 tablespoonfuls salad oil and a little pepper and salt; add to ½ cup white vinegar 4 tablespoonfuls water and pour it over the salad; mix it up well with 2 salad forks, sprinkle over a little cracked ice and serve at once. If ice is not handy the salad will have to be prepared without it, but it adds greatly to the crispness of the lettuce. If you rub a piece of garlic over the salad dish just before putting in the lettuce it will give the salad a fine flavor without really tasting of garlic. Asmall spoonful of sugar may be sprinkled over the lettuce if liked. Finely shaved onions may also be added.

855.**Lettuce Salad with Mayonaise.**— Prepare the salad as directed, put it into a salad dish and pour over a mayonaise dressing. Finely shaved onions may be added if liked.

856.**Lettuce Salad with Sweet Egg Sauce.**— Cut 2 ounces fat pork into very small dice and fry them a light brown; beat 2 or 3 eggs until very light and slowly add the pork, 3 or 4 tablespoonfuls vinegar and 2 tablespoonfuls sugar; mix this well together and pour it over the salad. This recipe will make a sufficient quantity to dress 3 heads of lettuce. More vinegar diluted with a little water may be added; also more or less sugar.

857.**Lettuce Salad with Syrup Sauce (North German art).** — Mix 1 tablespoonful flour in a small saucepan with a little cold water until all lumps are dissolved, add 1 cup boiling water and 253 stir and boil for a few minutes; add 1 tablespoonful butter and continue boiling for a few minutes longer; then transfer it to a bowl and set aside; when nearly cold add ½ cup syrup, apinch of salt, 2 tablespoonfuls vinegar and 1 tablespoonful sugar; set this sauce on ice or in a cool place for 5 minutes before serving; put the prepared lettuce in a dish and pour the sauce over it; sufficient for 4 large heads of lettuce. Salad prepared in this way is served in North Germany with German pancakes instead of butter. Fat pork cut fine and fried to a crisp may be used.

858.**Lettuce Salad with Cream Sauce (North German art).** — Prepare the lettuce as directed for 3 or 4 large heads of salad; take 1 pint thick, sour cream and add 3 or 4 tablespoonfuls sugar, apinch of salt and 1 or 2 tablespoonfuls vinegar; mix this together and pour over the salad; then serve at once.

859.**Lettuce Salad with Cream.** — Put the prepared lettuce in a dish and pour over some sweet cream to which a little sugar has been added. Some people add a little vinegar and a pinch of salt. Salad prepared with this sauce is often served with large German pancakes.

860.**Beet Salad.** — Wash and put ½ dozen beets in a saucepan with boiling water and cover and boil them till tender; when done put the beets into cold water, remove the skins and cut them while still warm into thin slices; also cut 1 medium sized onion into thin slices; put the beets and onion in alternate layers into a dish and sprinkle between 1 teaspoonful salt, ¼ teaspoonful pepper and 2 teaspoonfuls sugar; pour over an equal quantity of vinegar and water (enough to nearly cover the beets) and let them stand 1 hour before serving. Omit the onion if its flavor is not liked.

861.**Salad Macédoine.** — Take equal quantities of boiled white beans, boiled potatoes, celery roots, beets and string beans (the last 4 boiled in salt water) and cut into fine slices; put into a bowl 2 or 3 tablespoonfuls oil, vinegar and salt, pepper and some sugar; put in all the ingredients, add some finely chopped parsley and chervil and mix the whole together thoroughly; put the salad 254 into a dish and garnish with lettuce leaves. If the vinegar is too sharp dilute it with water.

862.**Salad à la russe.**— Boil 6 medium sized potatoes with the skins on, 2 beets and 3 celery roots; when cold remove the skins and cut them into small dice; also cut into dice 2 pickles and 1 dozen anchovies or 3 herrings previously soaked in water, freed from skins and bones and cut fine; add to this 2 tablespoonfuls capers, ½ cup grated horseradish and mix the whole with a fine mayonaise; put the salad on ice for 1 hour before serving; when ready to serve put the salad onto a round dish, pile up high in center and garnish with hard boiled eggs; chop fine the yolks and whites separate; also chop beets and green pickles fine, lay them in small clusters all over the salad and garnish the edge with green lettuce leaves or shaved pink and white horseradish. Pink horseradish is made by pouring a little cochineal over it and mixing well.

863.**Cucumber Salad.**— Select 3 medium sized cucumbers with small seeds, pare and cut a small piece from each end and lay the cucumbers in strongly salted ice water for 1 hour or longer; 10 minutes before serving take them out of water, wipe dry and cut on a board with a sharp knife into fine slices; put them into a salad dish, sprinkle over a little salt and pepper, pour over 2 tablespoonfuls salad oil and mix it with the cucumbers; then pour over ½ cup white vinegar to which a little water and a pinch of sugar has been added; if onions are liked cut a medium sized one into thin slices and put them between the cucumbers; some finely chopped parsley may also be added if the flavor is liked.

864.**Salad de laitue romaine.**— Take several heads of young, green lettuce; do not wash them; put them into a dish, add some coarsely cut chervil, tarragon and pimpernel and dress it either with salt, pepper, oil and vinegar or with mayonaise.

865.**Salad of Oyster Plant.**— Scrape and wash 2 bunches oyster plant and drop as you clean it into cold water to which 1 cup vinegar and 1 tablespoonful flour have been added; put a saucepan 255 over the fire with boiling water, add ½ cup vinegar, ½ tablespoonful flour wet with a little cold water, put in the oyster plant and boil till tender; when done drain in a colander and when cold cut it into lengths 2 inches long; arrange it nicely in a dish and pour a mayonaise over; or dress the oyster plant with oil, vinegar, pepper and salt.

866.**Asparagus Salad.**— Pare and cut into 2 inch lengths 1 bunch asparagus and boil it in salt water till tender; when done drain in a colander and when cold put the asparagus into a salad bowl; dress it either with mayonaise or pepper, salt, oil and vinegar.

867.**Carrot Salad.**— Scrape and wash ½ dozen medium sized carrots, put them in a saucepan over the fire with boiling water, add 1 tablespoonful sugar and boil till tender; when done take them out of water and set aside to cool; shortly before serving cut the carrots into fine slices, put them into a salad dish and pour over a mayonaise dressing; or dress the carrots with pepper, salt, oil and vinegar. If very large carrots are used first cut them in quarters and then into slices or dice.

868.**Carrot Salad with Asparagus.**— Prepare ½ dozen medium sized carrots the same as for Carrot Salad; when cold cut them into dice; boil the heads of 1 bunch asparagus in salt water till done, but not too soft; drain it in a colander and set with the carrots in the ice box for 1 hour; shortly before serving put the carrots and asparagus heads, in alternate layers, into a salad dish, pour over a mayonaise and garnish the dish with hard boiled eggs cut into slices and young lettuce leaves; sprinkle a few capers over the top.

869.**Carrot Salad with Onions.**— Prepare the carrots the same way as in foregoing recipe; cut 3 or 4 medium sized onions on a board with a sharp knife into slices as thin as wafers, put them in alternate layers with the carrots into a dish and pour over a mayonaise dressing; or dress with oil, vinegar, pepper and salt; add to vinegar a little water and sugar.

256

870.**Carrot Salad with Peas.**— Boil the carrots the same as for Carrot Salad and cut them into small dice; put 1 pint fresh green peas in a saucepan, cover with boiling water and add 2 teaspoonfuls sugar.

871.**Celery Root Salad.**— Boil ½ dozen celery roots; when done take them out of water and when cold pare and cut into quarters; cut each quarter into thin slices, put them into a salad dish and season with salt and pepper; add 2 or 3 tablespoonfuls salad oil and ½ cup vinegar; mix this well together and pour over ½ cup boiling water; or dress the salad with mayonaise and garnish with green lettuce.

872.**White Celery Salad.**— Take the white part of 1 or 2 bunches of celery and lay it for 1 hour in ice water; shortly before serving cut the celery into small pieces ½ inch in length, put it in a salad dish and pour over a mayonaise dressing; let it stand on ice for 15 minutes before serving. Some people use all of the celery except the leaves, but the salad is finer when made of the white part only.

873.**Cabbage Salad.**— Remove the outer leaves from a firm head of cabbage, shave it as fine as possible and put in ice water for 1 hour; before serving drain the cabbage in a colander, put it in a salad dish and mix with mayonaise; set it on ice until wanted; or dress the cabbage with oil, pepper, salt and vinegar; add to the latter before pouring it over the cabbage 1 spoonful sugar.

874.**Salad of Red Cabbage.**— Cut the cabbage as fine as possible, put it in a saucepan, pour over boiling water, cover and boil 3 minutes; drain in a colander and when cold dress it with oil, pepper, salt, asmall spoonful sugar and some vinegar; the latter should be diluted with water.

875.**Hot Slaw.**— Cut a small, firm head of cabbage as fine as possible and put it in a large bowl; place a saucepan with 1 cup vinegar, 1 tablespoonful butter and 1 teaspoonful sugar over the fire and let it come to a boil; then pour it over the cabbage and season with pepper and salt; at the same time put 1 egg with 1 cup 257 milk into another saucepan; beat these 2 ingredients together thoroughly and stir them over the fire till just about to boil; pour it over the cabbage and serve at once. Sweet cream may be used instead of milk.

876.**Radish Salad.**— Select 3 or 4 bunches nice, sound radishes, trim them neatly and lay for 1 hour in ice water; 10 minutes before serving wipe the radishes dry and cut them into fine slices; also cut 2 medium sized onions into fine slices like wafers; put a layer of radishes into a salad dish, sprinkle over a little salt and white pepper and put over a layer of onions with very little salt and white pepper; continue in this way in alternate layers until all is used; then pour over the whole a mayonaise dressing and garnish with green parsley leaves. The onions may be omitted if their flavor is not liked, but the salad is much finer with them. Instead of mayonaise the salad may be dressed with oil, pepper, salt and vinegar; the latter should be diluted with ⅓ water and a small spoonful sugar added to it before pouring over the salad.

877.**White Bean Salad.**— Wash and pick over 1 pint dry beans, put them over the fire in a saucepan, cover with cold water, add ½ teaspoonful carbonate of soda and boil 10 minutes; pour the beans into a colander and rinse with cold water; return them to saucepan again, cover with cold water, put a small piece of salt pork into the beans and slowly boil till the beans are tender; remove them from the fire and drain in a colander; when cold put them in a dish and season with pepper and a little salt; add 2 tablespoonfuls oil and 1 cup vinegar mixed with ½ cup water and a small spoonful sugar; shake all well together; add 2 tablespoonfuls finely chopped parsley and, if liked, afinely sliced onion; or dress the beans with mayonaise.

878.**String Bean Salad.**— Choose 1 quart young string beans, string and cut them into halves and boil in salted water until tender; when done drain them in a colander and when cold mix them with pepper, 2 or 3 tablespoonfuls oil, 1 cup vinegar mixed with a little sugar and ½ water and 1 finely cut onion; set the salad on ice 1 258 hour before serving. Butter bean salad is made the same as String Bean Salad, or dress with mayonaise.

879.**Crab Salad.**— Take 1 pint crab meat, sprinkle the juice of 1 lemon and 2 tablespoonfuls oil over it, mix well and set aside; shortly before serving put the crab meat on a salad dish with hard boiled eggs cut into small pieces; have some lettuce laying in ice water, drain it in a colander, shake dry in a napkin, cut through with a knife once or twice, mix some with the crab meat and use some for garnishing the dish; pour over the whole a mayonaise dressing the same as for Lobster Salad and serve.

880.**Pike Salad.**— Take a large sized fish, clean, wash, dry and cut it into 4 pieces; put it in salted boiling water and vinegar and add some onions and 1 bouquet; bring it to a boil quickly; then set the kettle aside and let it simmer till the fish is done; as soon as the fish is tender take it out and when cold remove skin and bones; cut each piece into 9 small pieces, lay them in a dish and pour over some sweet salad oil, tarragon, vinegar and sprinkle over a little salt and pepper; after the fish has laid 2 hours pour into a salad dish some mayonaise, lay in some fish pieces and pour over mayonaise; lay in the rest of fish and pour over the remaining mayonaise; garnish the dish with aspic or green lettuce and hard boiled eggs.

881.**Tomato and Potato Salad.**— Boil 6 large potatoes with the skins on; when cold pare off the skins and cut the potatoes into quarters; then cut each quarter into fine slices; lay 4 large, ripe tomatoes for ½ hour on ice; then cut them into small slices; cut 2 onions into fine slices, put them in alternate layers with the potatoes and tomatoes into a salad dish and sprinkle over each layer some pepper, salt and ½ tablespoonful oil; mix ¾ cup vinegar with 1 tablespoonful sugar and ¾ cup cold water, pour it over the salad and let it stand 15 minutes; then serve. The onions may be omitted and green peppers used instead of them.

882.**Tomato Salad with Lettuce.**— Lay 4 medium sized sound tomatoes for ½ hour on ice or in ice water; also the leaves of 259 2 large heads of lettuce; 10 minutes before serving wipe the tomatoes dry and cut them on a board with a very sharp knife into thin slices; shake the salad in a napkin; put into a salad dish first a layer of lettuce leaves and then a layer of tomatoes; continue in this way in alternate layers until all is used; pour over a mayonaise dressing and serve. 2 hard boiled eggs chopped fine may be sprinkled over the top.

883.**Tomato Salad.**— Lay ½ dozen sound, ripe tomatoes for 1½ hours on ice or in ice water; shortly before serving wipe the tomatoes dry and cut them on a board with a sharp knife into thin slices; also cut 2 medium sized onions into fine slices; put them in alternate layers with the tomatoes into a salad dish and sprinkle over each layer ¼ teaspoonful salt, ¼ teaspoonful sugar and half that quantity of white pepper; mix ¾ cup vinegar with ¾ cup water, pour it over the whole and serve at once. 2 tablespoonfuls oil may be added to the salad if liked, but many people object to it on tomato salad.

884.**Tomato Salad with Mayonaise.**— Lay 8 or 10 sound, ripe tomatoes for 1½ hours on top of ice; shortly before serving wipe the tomatoes dry and cut them on a board with a sharp knife into fine slices; put them into a salad dish, pour over a mayonaise dressing and sprinkle over the top 2 tablespoonfuls capers; serve as soon as made.

885.**Tomato and Cucumber Salad.**— Prepare the cucumbers the same as for Cucumber Salad, put them in alternate layers with the sliced tomatoes into a salad dish and dress with mayonaise.

886.**Tomato Pepper Salad.**— Lay 6 sound, ripe tomatoes for 1 hour on ice or in ice water; remove the seeds from 2 green peppers and throw the seeds away; lay the peppers for 1 hour in ice water; 15 minutes before serving wipe the tomatoes and peppers dry, cut the tomatoes into fine slices and cut the peppers into small pieces; put a layer of tomatoes in a salad dish and sprinkle over some of the finely cut peppers; then tomatoes again; 260 continue in this way until all is used; pour over a fine mayonaise and serve at once.

887.**Tomato Farce (à la Mayonaise).**— Prepare 2 heads of lettuce as directed and lay them in ice water; select 6 medium sized ripe tomatoes and lay them for 1 hour on ice; shortly before serving cut a thin slice off the blossom side of the tomatoes, scoop out the insides, chop fine with some white celery and the whites of 2 hard boiled eggs and mix with a few spoonfuls mayonaise; fill each tomato with this mixture; take 6 small dessert plates and put 1 tomato on each with 3 or 4 lettuce leaves around it; pour 1 tablespoonful mayonaise over each one and serve 1 to each person.

888.**Potato Salad.**— Wash and boil 12 medium sized potatoes with the skins on; when done drain off the water and pare off the skins; put into a deep bowl 2 finely cut onions, ½ cup white vinegar, 3 or 4 tablespoonfuls salad oil, 1 teaspoonful salt and ½ teaspoonful pepper; cut the potatoes while hot into fine slices and put them into the dish with the vinegar, oil and onions; pour over ½ cup boiling water; shake up the salad well in the bowl (do not stir it) and pour it into a salad dish; cover and let it stand for 1 hour; when ready to serve garnish the dish with finely cut beets and lettuce leaves. 2 tablespoonfuls finely chopped parsley may be mixed with the salad. Potato salad dressed with mayonaise is very nice.

889.**Potato Salad (another way).**— Wash and pare 12 medium sized potatoes and boil them in salted water till done; drain off the water and turn the potatoes into a dish; when cold cut them into slices; cut 2 good sized onions into fine slices as thin as a wafer; mix ¾ cup vinegar with ½ cup water and 1 teaspoonful sugar; put a layer of potatoes into a dish, then a layer of onions; sprinkle over some pepper and pour over 1 tablespoonful oil; put in another layer of potatoes and onions; continue in this way in alternate layers until all is used; pour the vinegar over the whole and cover and set in a cool place for 2 hours before serving.

261

890.**Potato Salad without Oil.—** Wash and boil the potatoes; when done drain off the water, pare off the skins and cut potatoes into slices; take for a soupplateful of sliced potatoes 2 finely cut onions and ¼ pound fat salt pork cut into small dice and fried a light brown; put potatoes, onions and pork into a deep bowl and season with pepper and salt; mix ½ cup vinegar with ½ cup boiling water and pour it over the potatoes; shake this well together and pour it into a salad dish; let it stand 1 hour or more before serving.

891.**Potato Salad without Onions.—** Wash the potatoes in several cold waters and boil them with the skins on; when done remove the skins and cut potatoes into slices; season them with salt and pepper, pour over an equal quantity of boiling water and vinegar and let them stand till cold; then add some sweet oil; mix it well and serve.

892.**Fine Potato Salad.—** Boil 10 medium sized potatoes with their skins on; when done remove the skins and set the potatoes aside to cool; stir the yolks of 2 eggs to a cream and slowly add ½ cup salad oil, 1 teaspoonful salt, ½ teaspoonful pepper, ½ cup white vinegar and 2 white onions chopped very fine; cut the potatoes into fine slices; put a layer of potatoes into a salad dish and pour over some of the sauce; then put in another layer of potatoes and sauce; continue in this way until all is used; then pour over ½ cup boiling water, cover and let it stand for 2 hours and then serve. The dish may be garnished with cresses or young lettuce leaves; or lettuce leaves and boiled beets cut into fancy shapes.

893.**Salad Endive.—** Take only young and fresh endive; remove the outer leaves, cut the endive into 1 inch pieces and wash and drain it; then dress it with oil, vinegar, pepper and salt, or with mayonaise.

894.**Beet and Cabbage Salad.** — Cut the white leaves of a savoy cabbage into shreds; remove the skins from 4 boiled beets and cut them into fine slices; chop a medium sized onion very fine; 262 put the cabbage and beets in alternate layers into a salad dish, sprinkle between the onions, some pepper and salt, pour over ½ pint vinegar, cover and let it stand 1 hour; then drain off the vinegar and add 4 tablespoonfuls sweet oil; mix it well with 2 salad forks, put in a salad dish and serve.

895.**Vegetable Salad.** — Cut with a tin tube some carrots and white turnips into small pieces and boil them separately in salted water; when done drain them in a colander; also boil small roses of cauliflower, green peas, beets and potatoes cut into small dice and some boiled string beans cut into small pieces; mix all the ingredients together and add 1 onion chopped very fine, some chopped parsley and chervil; add pepper, salt, oil and vinegar; put the salad into a salad dish and set it on ice for 1 hour.

896.**Beet and Potato Salad.** — Prepare a potato salad and mix it with half the same quantity of boiled beets cut into fine slices.

897.**Green Pepper Salad.** — Scoop out the insides of 2 green peppers and lay them for 1 hour in cold water; in the meantime prepare a mayonaise as follows: — Stir the yolks of 3 eggs to a cream and add by degrees 3 tablespoonfuls oil, stirring constantly; when this is well mixed add 3 tablespoonfuls vinegar; set this in a saucepan of hot water and stir and boil till the contents begin to thicken; then remove and stir until cold; slowly add 3 tablespoonfuls more oil, ¾ teaspoonful salt, 1½ teaspoonfuls sugar and 1 tablespoonful English mustard; shortly before serving add 1 tablespoonful condensed milk which is not sweet; chop fine 1 pound roast veal, lamb or boiled tongue, or 2 kinds of meat, 3 hard boiled eggs and 1 large onion; also chop the peppers very fine; mix all together, pour over the mayonaise, mix it well and set the salad on ice for 1 hour; in serving arrange it neatly on a salad dish, sprinkle over 1 tablespoonful capers and garnish with green lettuce leaves.

898.**Tripe Salad.** — Boil 2 pounds honeycomb tripe in salt water till tender; when done drain off the water, cut the tripe into 263 1 inch square pieces and mix it with mayonaise; put the salad into a dish and let it stand in a cool place for 1 hour before serving. Afew hard boiled eggs cut into slices may be added to this salad; also finely cut celery and lettuce leaves.

899.**Herring Salad.** — Lay 6 herrings for 24 hours in cold water, take them out, remove skins and bones, wipe dry with a towel and cut them into small square pieces; cut in the same manner 3 boiled beets, 2 white onions, 1 pound roasted veal, 3 sour apples and 5 small pickles; mix these ingredients together and prepare the following sauce: — Stir the herring milk to a cream and slowly add 1 cup salad oil, 2 teaspoonfuls sugar, the yolks of 4 eggs, ½ cup vinegar, 2 tablespoonfuls stewed cranberries, apinch of cayenne pepper and 4 tablespoonfuls French mustard; when all are well blended together mix the sauce with the herring and other ingredients and let it be ready 2 hours before serving; shortly before serving put the salad into a dish and garnish with small girkins, beets cut into fancy shapes, salted olives, hard boiled eggs and capers or mixed pickles.

900.**Salad à L'italienne.** — Soak 6 Dutch herrings for 12 hours in cold water; then take them out, remove the skins and bones and cut the meat into small long strips a little wider than a straw and ¾ inch in length; also cut 1 pound cold boiled beef tonge, 1 pound cold roasted veal, 3 greening apples and 6 small pickles; after the ingredients are all cut the same way mix them well together and mix them with a fine mayonaise; set the salad on ice for several hours; when ready to serve put the salad into a dish and garnish with finely chopped hard boiled eggs and salted olives; sprinkle over a few capers and serve.

901.**Chicken Salad.** — Select a plump 1-year old chicken for this; singe and draw it, wash in cold water and put the chicken in a kettle; cover with boiling water, add ½ tablespoonful salt and 2 onions; cover and boil slowly till tender; when done remove the kettle from the fire and let the chicken remain in the broth till cold; then take it out, remove the skin and bones and cut the meat 264 into small pieces; take the white part of 1 nice bunch celery and cut it very fine; add it to the finely cut chicken, pour over Mayonaise No. 1 and set it on ice for 2 hours before serving; when ready to serve put the salad into a salad dish and garnish with the small celery tops, which should lay for 1 hour in ice water; stick them all around in the salad and sprinkle ½ cup capers all over the salad. Some freshly grated cocoanut sprinkled over this salad is a great improvement; or garnish to taste.

902.**How to Boil Lobster.**— Select a good sized lobster, put it into a kettle of boiling water, head first, add a small handful salt and boil till the lobster has attained a bright red color, which will take from 20 to 30 minutes; when done take the lobster out and plunge it into cold water; let it lay in the cold water for 5 minutes; then take it out and when cold put the lobster away in a cool place till wanted.

903.**Lobster Salad.**— Split open the body and tail of a boiled lobster and crack the claws; pick out all the meat and cut it into pieces about ¾ inch in size; put the meat into a salad dish and pour over Mayonaise No. 1; let it stand in a cool place or on ice for ½ hour; then garnish the dish; lay a border of young lettuce leaves around the dish; lay over them some hard boiled eggs cut into quarters and sprinkle over the salad 1 spoonful capers. Canned lobster may be used for this salad. Another way is to cut white celery into small pieces, put it into a salad dish and mix well with a fine mayonaise; then add the lobster meat cut into small pieces and let the salad stand on ice or in a cool place for 1 hour before it is served; chop coarsely a few hard boiled eggs and sprinkle them over the salad.

904.**Salmon Salad.**— Select a nice piece of salmon weighing about 2 pounds; place a saucepan with boiling water over the fire and add a bunch of parsley with 2 bay leaves, 2 blades mace and a sprig of thyme; add 1 onion with 4 cloves stuck in it, 1 tablespoonful salt and ½ cup vinegar; when this boils put in the salmon and let it boil 3 minutes; then draw the kettle to side of stove and let it 265 simmer until tender; as soon as done remove the fish and set it in a cool place; when cold remove the bones and break the meat into pieces; put it into a salad dish, pile up high in the center, pour over a fine Mayonaise No. 1 and garnish the dish with young lettuce leaves or cresses; chop some hard boiled eggs and sprinkle them with a few capers over the salad. It is best to let the salad stand on ice for 1 hour before it is garnished and served. Canned salmon can be used instead of fresh salmon.

905.**Shrimp Salad.**— Extract the meat from some freshly boiled shrimp, put it into a dish, squeeze over some lemon juice, pour over a few spoonfuls fine oil and let it stand in a cool place for 1 hour; 1

hour before serving put the shrimp into a salad bowl, pour over a fine mayonaise (see Mayonaise) and garnish with cresses or lettuce leaves and hard boiled eggs cut into quarters.

906.**How to Boil Shrimp.**— Put the shrimp alive into the salted boiling water, allowing ¼ pound salt to 1 gallon water; boil them from 5 to 8 minutes; when they change color they are done; serve them with vinegar and oil.

907.**Halibut Salad.**— Put a piece of halibut into salted boiling water with ½ pint vinegar and add 1 or 2 onions, abunch of parsley, asprig of thyme, 1 bay leaf, 6 cloves and 2 blades mace tied together; bring it to a boil quickly; draw the kettle to side of stove and let the fish simmer until tender; when done take the fish out of the water and when cold cut it into 1 inch pieces; put the pieces into a dish high up in center and pour over it a mayonaise; garnish with green lettuce and hard boiled eggs.

908.**Oyster Salad.**— To make a salad for 6 persons take 2 dozen large oysters, put them with their liquor over the fire and let them boil 1 minute, but no longer; take them out with a skimmer and lay in a dish; when cold squeeze over the juice of 2 lemons and place the dish on ice for 1 hour; shortly before serving put the oysters into a salad dish, lay some young lettuce leaves between them and pour over Mayonaise No. 1; lay young lettuce leaves in a circle around the dish, put some hard boiled eggs cut into slices 266 between the lettuce and serve at once. Another way is to mix the oysters with finely cut white celery, dress them with the same mayonaise and ornament the salad with the tops of young celery; hard boiled chopped eggs may also be sprinkled over with 1 spoonful capers.

909.**Oyster and Chicken Salad.**— Remove the skin and bones from 1 cold, roasted chicken and cut the meat into pieces 1 inch in size; put it into a dish, sprinkle over a little salt, the juice of 2 lemons and pour over a few spoonfuls fine salad oil; then place the dish on ice; in the meantime scald 1½ dozen large oysters in their own liquor, take them out and put the oysters in a dish with some cracked ice; have prepared 2 quarts sour jelly (aspic) and pour a few spoonfuls of it onto a large, shallow tin pan; when firm trim the oysters so that there is nothing left but the eye; lay them over the jelly (not too close together), pour over a little more cold jelly and when firm pour over sufficient cold jelly to entirely cover the oysters; let it stand in a cool place till firm; 10 minutes before serving wipe the chicken meat dry with a napkin; pour some fine mayonaise into a salad dish, lay over a layer of the chicken meat and cover with mayonaise; continue in this way till all is used; cover the whole with mayonaise in such a way that none of the chicken is seen; then lay a border of cresses around it; cut the oysters into rounds with a fluted cutter a little larger than the oysters, lay them on the cresses and serve. Lettuce may be used instead of cresses.

910.**Tomato Jelly.** — Stew for ½ hour 1 can tomatoes with 1 teaspoonful salt, 1 teaspoonful sugar, as much cayenne pepper as you can hold on the point of a knife and 2 tablespoonfuls vinegar; then press them through a sieve; in the meantime soak 1 ounce gelatine in ½ cup cold water for 15 minutes, add it to the tomatoes, put over the fire and stir till the gelatine is dissolved; then strain through a flannel jelly bag; fill the jelly into small patty forms and set them in a cool place till firm.

911.**Tomato Jelly Salad.** — Prepare a tomato jelly the same as in foregoing recipe; turn it out of the small forms, lay 267 into a salad dish, stick small pieces of white celery into each one, put a border of young lettuce leaves around it, pour over a mayonaise and serve at once. Tomato jelly may be made in one large form and when hard chopped coarsely and used for garnishing dishes of cold meats or salads.

912.**Egg Salad.** — Put ½ dozen eggs in a saucepan, cover with cold water and boil them 10 minutes; transfer the eggs to cold water and let them lay till cooled off; when cold remove the shells and cut the eggs into quarters; put them into a salad dish with young lettuce leaves, pour over a mayonaise dressing and garnish with lettuce leaves.

913.**Eggs with Mayonaise.** — Boil ½ dozen eggs 10 minutes; then transfer them to a pan of cold water and when cold remove the shells; take 6 small plates, put 2 lettuce leaves on each plate and put an egg in the center of the 2 leaves in such a way that the leaves stand round the egg like a tulip; pour over each egg 1 tablespoonful mayonaise and sprinkle over a few capers; serve a plate to each person.

914.**Onion Salad.** — Take 2 large Bermuda or California onions, peel and cut them with a sharp knife into fine slices, put a layer of the slices into a salad dish and pour over some fine mayonaise; then put over some cresses and pour over more mayonaise; continue in this way until all is used; cover with mayonaise, lay some cresses in a circle round the dish and let it stand on ice for 10 minutes; then serve.

915.**Alligator Pear Salad.** — Take 2 alligator pears, cut them into slices and put them into a salad dish; remove the shells from 4 hard boiled eggs, break the yolks into small pieces and sprinkle them over the sliced pears; cut the whites into fine strips, lay them in a circle round the dish close to the pears, pour over a fine mayonaise and lay a border of tender lettuce leaves round the edge of dish.

916.**Jerusalem Artichoke Salad.** — Scrape the artichokes carefully and drop them into vinegar and water; mix ½ 268 tablespoonful flour with a little cold water, stir it into a quart of boiling water and add 1 cup vinegar; as soon as this boils put in the artichokes and boil them till done, but not too tender; when done remove them from the water and set in a cool place; when cold cut the artichokes into pieces, put them into a salad dish, pour over a mayonaise, set some shrimp around the salad and set the dish on ice for 1 hour; when ready to serve lay a border of lettuce leaves round the edge of dish.

917.**Sour Jelly (Aspic).** — Soak 2 ounces gelatine in ½ pint cold water 15 minutes; then put it over the fire with 1 quart good meat stock and sufficient vinegar to give it a nice sour taste; add a few cloves, 2 blades mace and 1 bay leaf; stir this over the fire till the gelatine is dissolved; beat the whites of 2 eggs till light and add the juice of 1 lemon and a little cold water; stir it with an egg beater into the jelly and stir and boil for a few minutes; then draw the saucepan to side of stove and let it stand 5 minutes; then strain through a jelly bag; or turn a chair upside down on a kitchen table; then take a square piece of unbleached muslin and tie a corner over each of the upturned legs of the chair; set a bowl underneath and pour the jelly onto the cloth a little at a time and keep the saucepan on the side of stove, to keep the jelly warm. If meat stock is not handy dissolve 2 teaspoonfuls Liebig's beef extract in 1 quart boiling water and use it instead of meat stock. Another way is to boil 4 calves' feet till they fall apart; then strain off the liquor, set it aside and when cold remove all the fat; boil the liquor down to 2 quarts; then beat the whites of 4 eggs to a froth and add the juice of 1 lemon and a little water; add to the broth sufficient white vinegar to give it a nice sour taste; also add a little salt, some whole pepper corns, afew blades mace, 4 cloves and 1 bay leaf; stir in the beaten whites, continue stirring, let the contents boil for a few minutes and let it stand 5 minutes; then draw to side of stove, let it stand 5 minutes and strain through a flannel jelly bag. Pigs' feet or the skin of fresh pork may be used instead of calves' feet. Sour jelly is used for garnishing dishes of meat and salads. It is either chopped with a knife or put into small 269 fancy forms and when firm turned out and laid around the dishes with cresses, lettuce or celery between. If the jelly is not dark enough add a little sugar color (see Sugar Color). If the jelly is white it may be colored green with green spinach color or pink with cochineal.

918.**Garnishing.**— The articles which are mostly used in garnishing are:—Lettuce, cresses and hard boiled eggs (either cut into slices or quarters or chopped fine, the yolks and whites separately, and laid alternately in small clusters all over the salad); or cut green pickles in slices and lay them in a circle around the salad with small clusters of finely chopped beets and chopped eggs; small girkins, capers, olives and very small, white pickled onions are also used for garnishing. Another way is to cut boiled carrots, white turnips and beets into fancy shapes, such as half moons, stars, leaves and roses, with a vegetable cutter; anchovies are also largely used in garnishing. They are freed from skins and bones and then rolled up and laid in a circle around the dish with small white onions, pink horseradish and olives or girkins.

919.**Horseradish for Garnishing.**— Remove the outside black skin from a large root of horseradish and wash it clean; then shave it off with a knife in long narrow strips so they curl up; color ½ the shavings with prepared cochineal and leave the other ½ white; then use for decorating dishes of meats or salads by laying it in small clusters around the dish.

920.**Cocoanut for Garnishing Salads.**— Grate cocoanut and sprinkle it over the top of salad. Especially nice over chicken, shrimp and fish salads; also on potato, tomato and egg salads. Grated cocoanut lends a handsome appearance to any salad.

ICES AND GLAZES.

921.**How to Use Icing.**— Over large cakes pour the icing by spoonfuls near center on top of cake and spread it with a broad-bladed knife dipped in cold water all over the cake as smoothly as possible; set it in a cool oven for a few minutes, then in a dry, airy place, free from dust, to dry. Some icing does not need to be put in the oven, as it dries immediately, as will be seen from the directions given in following recipes. Small cakes are dipped into the icing or into glaze and then laid on paper or tins to dry. If the cake is to be ornamented make a paper funnel as follows:—Take a piece of brown paper, not too thick, or white tea paper 12 inches square and cut it through on the bias in two pieces from one corner to the other; take one piece in your right hand, the bias side from you, roll with the left hand, the bias side towards you, and form the paper into a funnel; bend the end where it closes near the top over to the inside, clip a small piece from the end of funnel with a scissors and slip a small tube inside it to the end opening; then put in the icing and bend the top of funnel in all around the same way as ¼ pound tea is put up in those small funnel-shaped bags; next press the icing down towards the end and commence to squirt it onto the cake. The cake may be ornamented with a border and a harp in the center, or an anchor or any kind of a pattern that may be desired. Flowers and leaves may be bought at any confectionery and pasted on with a little icing.

922.**White Icing.**— Sift ½ pound powdered sugar into a bowl, add the whites of 2 eggs and stir 20 minutes; add a few drops lemon juice while the stirring is going on; drop a little icing onto paper; if the icing stands without running it is stiff enough; if it shows the least tendency to run more sugar must be added. This icing is used for ornamenting cake and serves as a kind of paste to stick flowers and leaves onto top of cake.

923.**Clear Icing.** — Sift ½ pound powdered sugar into a bowl, add the whites of 2 eggs and stir for 5 minutes; add a few drops lemon juice and stir 5 minutes longer; then spread it over the cake; set the cake for 2 or 3 minutes in a cool oven, take out and let it dry for a few hours in a dry place which is free from dust.

924.**White Icing with Wine or Liquor.** — Mix ½ pound sifted powdered sugar with the white of 1 egg and add 3 271 tablespoonfuls maraschino, Madeira or sherry wine, kirsch, rum or brandy; before this icing is put on cover the cake with a layer of jelly; then put the icing over it and set the cake for 1 minute in the oven; then set it in a dry place which is free from dust to dry. To make pink icing add a few drops prepared cochineal or strawberry syrup. Yellow icing is made by adding prepared saffron. Essence of lemon or the grated rind of 1 orange may be used instead of vanilla flavoring.

925.**Almond Icing.** — Pound 3 ounces blanched almonds with the white of 1 egg in a wedgewood mortar to a paste, mix them with ¼ pound powdered sugar, the white of ½ egg and ½ teaspoonful vanilla extract and stir for 5 minutes; dip small pieces of cake into the icing; pour and spread it over whole cakes with a broad-bladed knife. Hazel nuts and walnuts may be used the same way as almonds, as can also pistachio nuts. To the latter add a few drops spinach green.

926.**Fruit Icing.** — Mix ½ pound sifted powdered sugar with the white of 1 egg and add 3 tablespoonfuls fruit juice — either raspberry, strawberry, currant, pineapple or peach; if lemon or orange juice is used add a little grated rind; spread the icing over the cake and set it for a few minutes in a cool oven; then set it in a dry place which is free from dust to dry.

927.**Sugar Glaze.** — Mix 1 cup powdered sugar with 2 tablespoonfuls water and put it over the fire to get lukewarm; pour over the cake and let it dry, which will take but a few minutes; dip small pieces of cake into it. Glazes of raspberries, strawberries, pineapples, peaches, wine, maraschino or rum are made the same way. Omit the water and use 2 tablespoonfuls fruit juice or wine, whichever kind is wanted.

928.**Maraschino Glaze.** — Mix 1 cup sifted powdered sugar with 1 tablespoonful water and 1 tablespoonful maraschino, let it get warm on the fire and pour while warm over the cake. It will 272 get hard in a few minutes. Rum glaze is made the same as Maraschino Glaze.

929.**Orange Glaze.**— Mix 2 tablespoonfuls orange juice with 1 cup powdered sugar, add a little grated rind, set over a slow fire to get lukewarm and use it at once.

930.**Lemon Glaze.**— Stir 1 cup sifted powdered sugar with 1 tablespoonful lemon juice, 1 tablespoonful water and a little grated rind; let it get lukewarm; then spread it over the cake and set in a dry place which is free from dust to dry.

931.**Coffee Glaze.**— Mix 1 cup powdered sugar with 2 tablespoonfuls strong coffee, let it get lukewarm and use at once.

932.**Wine Glaze.**— Boil 1 cup sugar with ½ cup water until it forms a thread between 2 fingers; remove it from fire, add 2 tablespoonfuls sherry or Madeira wine and stir for 1 or 2 minutes; then quickly pour it over the cake and let it stand in a dry place which is free from dust to harden.

933.**Boiled Cinnamon Glaze.**— Boil 1 cup sugar with ½ cup water and 1 teaspoonful powdered cinnamon until it forms a thread between 2 fingers; remove it from fire, stir for a few minutes and use at once.

934.**Chocolate Glaze.**— Melt 2 tablespoonfuls grated chocolate in the oven and mix it with 2 tablespoonfuls sugar syrup; mix 1 cup sugar with 1 tablespoonful water, add the chocolate, set it over a slow fire to get lukewarm and use at once.

935.**Cinnamon Glaze.**— Mix 1 cup sifted powdered sugar with 1 teaspoonful cinnamon, add 2 tablespoonfuls water, set it over a slow fire, stir until lukewarm and use at once.

936.**Cold Sugar Glaze.**— Mix 1 cup sifted powdered sugar with 2 tablespoonfuls cold water and add 1 teaspoonful lemon or vanilla extract; spread glaze over the cake and set it for 1 or 2 273 minutes in a cool oven to obtain a glaze; then remove and set in a dry place which is free from dust to dry. NOTE.—Instead of water any kind of fruit juice syrup, wine, rum or Cognac may be used. If lemon juice is used take ½ water, ½ juice and a little grated rind. For orange use a little rum and 2 tablespoonfuls orange juice. For coffee use instead of water 2 spoonfuls strong coffee. For chocolate stir in 2 tablespoonfuls grated chocolate or cocoa the same way, or melt the chocolate in the oven and then add it to the sugar.

937.**Boiled Chocolate Glaze.**— Place a small saucepan with ½ pound sugar, ½ pound grated chocolate and ½ pint water over the fire and stir and boil till it forms a thread between 2 fingers; remove from fire and stir until a thin skin forms on top of glaze; then use it at once; spread it evenly all over the cake and set for a few minutes in a cool oven. If the glaze should become too cold before it is all used return it to the fire and repeat again. The glaze when done should shine like a mirror.

938.**Transparent Glaze.**— Boil 1 cup sugar with ¾ cup water until it forms a thread between 2 fingers; remove it from fire, stir for a few minutes and then quickly use; dip small pieces of cake into the glaze, pour over large pieces and spread it apart; let it dry in an airy room which is free from dust.

939.**Rose Glaze.**— Boil 1 cup sugar with ½ cup water till it forms a thread between 2 fingers; remove at once and add 2 tablespoonfuls rose water and a little prepared cochineal, to color it to a delicate pink; stir for a minute and then pour it over the cake. Small cakes may be dipped into the glaze and set in a dry place which is free from dust to dry.

940.**Spinning Sugar.**— Put ¾ pound loaf sugar in a small copper kettle, add sufficient cold water to half cover the sugar and stir until it is melted; then place the kettle over a strong fire and boil the sugar to a crack (the 6th grade); add a few drops vinegar, 274 remove the kettle, dip it for a few minutes into cold water and let it cool off a little; if the sugar is spun when too hot the threads will be too thin and lumps will form; then place the kettle in a pan of hot water, or on the side of stove, to keep the sugar warm; take a large knife in the left hand and hold it out straight before you; take a silver spoon in the right hand, dip it into the sugar without touching the bottom of kettle and let some of the sugar run off from the spoon; then spin long threads back and forth over the knife from right to left; after a considerable amount of sugar is spun in this way take it from the knife, lay on clean paper and spin the rest in like manner; when all is spun form the sugar into pompoms, garlands, bouquets, etc. Half the sugar may be colored with cochineal to a delicate pink. The sugar should be spun in a place free from draughts and in clear and dry weather. This sugar is used for decorating and trimming dishes.

941.**Boiling Sugar.** — Put 1 pound sugar into a kettle with ½ pint water and let it stand for a few minutes; then put it over the fire to boil; while the boiling is going on dip a small brush into cold water and wipe off the sides and edge of kettle; the different grades the sugar goes through in boiling are as follows: — 1st grade, broad run; 2d grade, small pearl; 3d grade, large pearl; 4th grade, the small blubber; 5th grade, the large blubber; 6th grade, to a crack; 7th grade, caramel; boil the sugar for a few minutes and dip the point of a spoon into it; if the sugar falls in large drops from the spoon it has reached the 1st grade; continue the boiling for a few minutes longer; dip your first finger into it and press the finger against the thumb; then open the fingers and if the sugar forms a thread between the 2 fingers it has reached the 2d grade; after boiling a little longer dip in a spoon and if a pearl hangs onto a long thread of the spoon the sugar has reached the 3d grade; after a few minutes longer boiling take a little in a spoon, blow it and if the sugar falls from the spoon in blubbers it has reached the 4th grade; after a few minutes longer take a little of the sugar between your fingers and quickly dip into cold water; if the sugar can be formed into a ball it has reached the 5th grade; after a few minutes longer 275 dip the finger into the sugar and then quickly into cold water; if the sugar can be broken it has reached the 6th grade; then set the saucepan in cold water; if it boils a few minutes longer it will have reached the 7th grade, or caramel. The principal care in boiling sugar is to use the exact amount of water. With too little water the sugar will curdle before it has boiled enough. If too much water is used the sugar will have to boil too long and will turn yellow. It should boil quickly and only for a short time. It will then stay white.

277

APPENDIX.

CAKES.

Cheese Torte.— One pound fresh pot cheese, ½ pint sour cream, 1½ ounce sweet almonds, 1½ ounce bitter almonds, 1 cup seedless raisins, 2 tablespoonfuls flour, 1 cup sugar, 1 tablespoonful butter, 6 eggs, ¼ teaspoonful salt. Blanche and grind the almonds fine or grate them on a nutmeg grater; stir sugar, butter, and yolks to a cream, add all the ingredients, and last the beaten whites; mix well and set aside till following paste is made: Sift 1 pint of flour with 1 teaspoonful baking powder into a bowl, add 2 tablespoonfuls sugar, ½ teaspoonful salt, and 1 tablespoonful butter. Rub the butter fine in the flour, add 1 cup of milk and 1 egg, mix together into a firm dough, work it lightly on a board till it does not stick to the hands, then roll it out thin. Butter a large cheesecake pan, dust it with flour, and line the pan with the dough; pour in the cheese preparation, and bake in medium-hot oven till nearly done. In the meantime stir the yolks of 2 eggs with 3 tablespoonfuls granulated sugar to a cream, add 1 teaspoonful lemon juice and little grated rind, add the 2 beaten whites, and stir the whole 10 minutes; add last 3 level tablespoonfuls flour. When the cake is firm to the touch remove it from the oven, pour over this last mixture, and bake till done. Serve cold, dusted with sugar. The bitter almonds may be omitted if objected to, and the cake pan may be lined with puff paste or fine pie crust.

Pistachio Torte.— Four ounces almond paste, 4 ounces ground pistachio nuts or almonds, the yolks of 16 eggs, the whites of 8 eggs, 3 ounces flour, ½ pound sugar, ½ teaspoonful extract of pistachio; rub the almond paste with the white of 1 egg fine; add the 16 278 yolks and sugar; stir 15 minutes, then add the ground pistachio nuts or almonds; continue the stirring 10 minutes; add the extract; beat the whites to a stiff froth; add the yolk mixture to the whites while beating constantly; beat the whole together 5 minutes; add the sifted flour; stir the flour in lightly; butter 2 large deep jelly-cake tins and dust them with the flour; fill in the mixture and bake in a slow oven. FILLING: Boil ½ cup of sugar, with ½ cup water to a caramel, then add it slowly to the beaten whites of 2 eggs; beat until cold; add ½ teaspoonful vanilla sugar, ½ cupful fine-cut candied pineapples, ½ cupful fine-chopped pistachio nuts or almonds; spread this filling over one layer, put on the remaining layer. Ice the cake with pistachio icing made as follows: Mix 1½ cup sifted powdered sugar with 3 tablespoonfuls boiling water and a little green coloring and pistachio flavoring; pour over the cake and let stand till firm.

Kugelhupf, also called **Bunt Kuchen.**— One pound flour, 10 ounces butter, 2 yeast cakes, 6 whole eggs and 6 yolks, ¾ cup of sugar, ½ teaspoonful salt, 1 cupful seeded raisins; break the yeast into a small bowl, add ½ cupful lukewarm milk and 1 teaspoonful sugar; let it stand till the yeast rises to the surface, then add ½ cupful of flour, mix to a stiff batter; cover and let stand till it is a light sponge. In the meantime stir butter and sugar to a cream; add the yolks one at a time, then alternately a little flour and 1 whole egg, till all are used; beat this with the right hand 10 minutes; add the raisins, and last the sponge; continue to beat 5 minutes; butter a large ribbed form with tube in the center; dust with powdered sugar, pour in the cake mixture; set the form in a warm place till the contents has risen to double its size; then place the form on a tin with salt and bake in a medium-hot oven about 45 minutes. When done, take it out of the oven, let it stand a few minutes, then turn the cake out of the form, dust it with sugar, and serve when cold.

Jelly Roll.— One half pound sugar, 9 ounces flour, 1 teaspoonful baking powder, 1 gill of milk (½ cupful), 3 eggs, ½ teaspoonful lemon extract; sift flour, sugar, and powder into a bowl and make a hollow in the center; put in milk, eggs, and lemon extract, mix all together; butter a large shallow tin pan, cover with paper, spread 279 on the mixture thin and evenly, bake in slow oven. When done, remove the pan, let it stand a few minutes to cool off a little, then turn it upside down on a clean piece of paper, remove the paper carefully from the cake which has laid in the pan, spread some currant jelly over the surface, roll the cake up like a music roll, let it lie rolled in the paper till cold. This preparation is also nice for lady fingers.

Election Cake.— One and a half pint lukewarm milk, 1 pound sugar, ½ pound lard, ½ pound butter, 2 pounds flour, the whites of 4 eggs, 1 pound citron, 1 pound seeded raisins, 1 teaspoonful mace, ½ cupful rum, 3 yeast cakes. Break the yeast into a cup of lukewarm milk, add 1 teaspoonful sugar, set the cup in a warm place till the yeast rises to the surface, put 1 cupful flour into a bowl, add the yeast, mix into a stiff batter, cover, and set in a warm place to rise till the sponge is very light. In the meantime stir butter, lard, and sugar to a cream, add the mace, then alternately milk and flour, then the fruit and rum, and last the 4 whites beaten to a stiff froth; beat the whole with the hand 10 minutes, then add the sponge; continue to beat a few minutes longer, cover, and set it in a warm place to rise till double its size; butter and dust with flour a large round cake pan, pour in the cake mixture, and bake in a medium-hot oven till done. When cold, ice the cake with rum icing.

Small Sponge Cake.— Three eggs, ½ cup granulated sugar, ½ cup flour, the grated rind of ½ lemon, and a little lemon juice. Stir the 3 yolks with the sugar 15 minutes, then add the lemon; beat the whites to a stiff froth, add them to the yolks, and beat till the sugar is all dissolved, which will take about 10 minutes, then sift in the flour, stir the flour in lightly; butter and dust with flour a small round pan, pour in the mixture, and bake in a slow oven.

Apple Ringlets.— Peel, core, and cut into thick slices 4 large tart apples, mix ½ cup of flour with ½ teaspoonful baking powder, ½ teaspoonful butter, and ½ teaspoonful salt. Break 1 egg into ½ cup of cold water, beat until it foams, add the water and egg to the flour, and mix into a batter. Melt 1 tablespoonful lard and butter in a frying pan, dip the apple slices into the batter, put them into the frying pan, not too many at once, and fry light brown on 280 both sides, keeping the pan covered while the frying is going on. Serve dusted with sugar.

Baking-Powder Rolls.— One pint flour, 1 heaping teaspoonful baking powder, ½ tablespoonful butter, ½ teaspoonful salt, ½ pint milk, 1 egg, 1 teaspoonful sugar. Sift flour, salt, sugar, and powder into a bowl, add the butter and rub it fine with the flour, mix the egg and milk together, pour a little of the egg milk into a cup, add the remaining to the flour, mix all together with a knife into a firm dough, turn it on to a floured board, and work it together to smooth the dough, roll it out ¼ inch in thickness, then cut it into rounds, brush them over with a little melted butter, fold them over and set them on a buttered tin, brush the rolls over with the egg milk which was set aside, and bake in a quick oven. Agood plan is to keep the rolls covered with buttered paper the first 10 minutes while baking.

Waffles.— Four ounces butter, 6 eggs, 4 ounces flour, 1 tablespoonful sugar, apinch of mace, and a little grated lemon rind, ½ pint whipped cream. Stir the butter to a cream, add alternately 1 yolk, then a little flour until both are used, add the sugar, the mace, and lemon, then the whipped cream, and last 5 whites whipped to a stiff froth; rub a waffle iron with fork, pour 1 tablespoonful batter into each compartment, and bake the waffles light brown on both sides. Serve dusted with sugar.

Gateau à la Weckesser.— Half pound of granulated sugar, 13 yolks, 9 whites, ½ pound flour, and the rind and juice of 1 lemon. Stir sugar and yolks for 25 minutes by the clock, then beat the whites to a stiff froth; add the yolks and sugar slowly to the white while beating constantly, add the lemon, continue the beating 5 minutes, then add the sifted flour, stir in lightly; butter a large round pan and dust it with flour, pour in the batter, place the cake on a pan of salt, and bake in a slow oven. When done, turn it on to a board, which should be dusted with powdered sugar; let it lie till cold, then spread a layer of pineapple marmalade over the cake, ice it with white sugar glaze, and decorate the cake with candied fruit, of plums, apricots, and cherries. The fruit must be cut into small slices and the cherries in small dice.

281

Lady Cake.— Six ounces butter, ½ pound sugar, 1 pound flour, 10 whites of eggs, ½ pint whipped cream, 1 teaspoonful baking powder, 1 teaspoonful vanilla sugar. Stir butter and sugar to a cream until white and frothy, add the vanilla sugar, sift flour and baking powder together, beat ½ pint cream until stiff; beat also the 10 whites to a stiff froth, then add alternately the three ingredients to the creamed butter and sugar; butter and dust with flour a large round pan, pour in the mixture, and bake in a medium-hot oven. When done and cold, ice it with clear icing.

Denmark Cake.— Two pounds flour, 2 teaspoonfuls baking powder, 1¾ pound sugar, 1 pound butter, 2 pounds raisins, 8 eggs, 1 pint sweet milk, ½ pint wine, 3 tablespoonfuls allspice, 3 tablespoonfuls cinnamon, 4 nutmegs. Sift flour and baking powder together, stir butter and sugar to a cream, add gradually the yolks and spice, then alternately milk, flour, and wine, last the fruit. Bake in a large well-buttered pan in medium-hot oven.

Stullen with Baking Powder.— One pound flour sifted with ½ teaspoonful salt, 2 teaspoonfuls baking powder, ½ cup of butter, ½ cup of sugar, and 2 eggs, the fine-chopped peel of ½ lemon, 1 cupful seeded raisins, ½ cupful fine-cut citron, ¾ cup of milk. Rub flour and butter together, add sugar, salt, milk, and eggs, mix all together; add last the fruit, turn the dough on to a floured board and work it a little to smooth the dough, then roll it out 1 inch in thickness, fold it over and lay the cake in a buttered pan, giving it the shape of a half moon; brush over with beaten egg and bake in medium-hot oven.

Wine Baba.— One pound flour, 2 teaspoonfuls baking powder, 1 teaspoonful salt, 4 ounces butter, 3 tablespoonfuls sugar, the fine-chopped peel of 1 lemon, 4 eggs, ½ pint milk, ½ cup currants, ½ cup seeded raisins, ½ cup fine-cut citron. Sift flour and baking powder together, stir butter and sugar to a cream, add by degrees the yolks and lemon, then alternately flour and milk, next add the fruit, and last the beaten whites. Fill the mixture into a buttered form with a tube in the center, place it in a medium-hot oven, and bake about 40 minutes or till done. In the meantime boil 1 cup sugar with ½ cup water 5 minutes, remove from fire, add 1 gill 282 sherry wine and a few tablespoonfuls of raspberry syrup. When the cake is done turn it on to a sieve, place the sieve on a large plate, and pour the syrup by spoonfuls over the cake; pour that which runs below in the plate over the cake again.

Bunt Kuchen with Baking Powder.— One pint flour, 1 heaping teaspoonful baking powder, ½ teaspoonful salt, 3 tablespoonfuls sugar, 2 ounces butter, 2 eggs, ½ pint milk, ½ cup currants, ½ cup stoned raisins, the grated rind of ½ lemon. Sift flour and baking powder together, add the sugar, salt, and butter, rub the butter fine in the flour, mix the yolks with the milk and add them to the flour, mix all into a dough, add the fruit, and last the beaten whites. Fill the mixture into a round buttered form with a tube in center and bake about 35 minutes; cover the first 20 minutes with buttered paper.

Emelines. — *First Part*: The whites of 6 eggs, 1 cup powdered sugar, 1 cup flour sifted with 1 teaspoonful baking powder, ½ teaspoonful lemon extract, 1 tablespoonful melted butter, measured after it is melted, 2 tablespoonfuls milk. Beat the whites till stiff, then add gradually the sugar, butter, lemon extract, and milk, and last the flour. Butter a long shallow tin pan (13 inches long, 9 inches wide, and 1 inch deep), dust it with flour, pour in the mixture, smooth it even with a knife, and bake in a medium-hot oven.

Second Part: The yolks of 6 eggs, 1 cup powdered sugar, 1 cup flour sifted with 1 teaspoonful baking powder, 2 tablespoonfuls milk, 1 tablespoonful butter, ½ teaspoonful lemon extract. Stir butter, sugar, and yolks to a cream, add lemon extract, flour, and milk; bake the same as the first part. When the cakes are nearly cold lay them over one another with a layer of jelly between, then cut it into fancy shapes like half moons, small rounds, and squares; glaze them with sugar glaze No. 927. In place of jelly, cream or any other filling may be taken. The two cakes may be spread separately with jelly, then rolled up like jelly rolls.

Sand Wafers. — Stir 4 ounces butter with 6 tablespoonfuls sugar till light and creamy, add gradually 3 eggs, the grated rind of ½ lemon, stir 15 minutes, add last 7 ounces sifted flour. Fill the preparation in a pastry bag, butter and dust with flour some large 283 shallow tin pans, press small cakes from the bag on to the pans the size of a 50-cent piece, and bake in medium-hot oven. When done and cold, glaze them with fruit glaze or leave them plain.

Cream S. — Stir ½ pound butter with 6 ounces sugar to a cream, add 1 teaspoonful vanilla sugar and 4 eggs, stir 10 minutes; add last ½ pound of sifted flour. Put the preparation into a kiss syringe and press small cakes in the shape of an S into buttered and floured pans, bake in medium-hot oven, and when cold glaze them with vanilla glaze.

Aniseed Wafers. — Rub some shallow tin pans with wax, place ½ pound sugar and 4 whole eggs in a bowl, set the bowl into a pan of hot water, beat the contents of bowl with an egg beater 15 minutes, then remove and beat till cold; add 1 teaspoonful well-cleaned aniseed and ½ pound sifted flour, fill the mixture into a pastry bag and press small cakes on to the waxed tins, cover and let them stand till next day, when the little cakes have obtained a crust, then bake them in slow oven.

Cinnamon Sticks.— Four ounces of almond paste, the white of 1 egg, 4 ounces powdered sugar, 1 teaspoonful cinnamon. Mix all together, put on to a floured board; roll it out ⅛ inch thick, cut the paste into 3 long strips 3 fingers wide, spread over an icing made as follows: Mix the whites of 2 eggs with ½ pound powdered sugar, add ½ teaspoonful cinnamon and a few drops of lemon juice, stir 5 minutes, then cut the strips into small sticks a finger wide, lay them on to buttered tins, and bake in a slow oven.

Meringue Shells.— Beat the whites of 4 eggs to a stiff froth, add ½ pound granulated sugar and a little vanilla sugar, stir the sugar in lightly; fill this into a pastry bag or a paper funnel; press small portions on to a double-folded paper, dust thickly with sugar, lay the paper in a pan, and bake in a slow oven. When done and cold, remove them from the paper, press the soft bottoms into shape to form a shell, and serve filled with whipped sweetened cream or ice cream.

Kisses.— Five ounces whites of egg, 1 pound granulated sugar, 1 cup water, 4 ounces powdered sugar. Put the granulated sugar 284 and water in a saucepan, stir until the sugar is dissolved, then place the saucepan over the fire, and boil to a crack; have the whites beaten to a stiff froth, and add slowly the hot sugar while beating constantly with an egg beater, then beat until cold. This may then be used as it is, and if wanted very stiff, add the 4 ounces powdered sugar; stir it through the mixture lightly, then put the meringue in a kiss syringe, dust some paper with powdered sugar, press the mixture on to the paper in any shape desired. For shells it may be put on with a spoon, dust them with sugar, and bake in a slow oven. The oven may be left open part of the time. In place of paper, rub some tin pans with butter, then rub off all the butter with paper, and dust them with powdered sugar, then put the kisses on them.

Banana Cake.— Three bananas, 1 cup of currant jelly, ½ pint of whipped cream, 3 ounces butter, ¾ cup sugar, 1½ cups of flour, 1 teaspoonful baking powder, the whites of 3 eggs beaten to a stiff froth, ½ cup milk, the juice and grated rind of ½ lemon. Sift flour and baking powder together, stir butter and sugar to a cream, add the lemon, then alternately flour, milk, and the white of egg; butter 2 jelly tins of medium size, dust them with flour, divide the cake mixture evenly in the tins, and bake in a medium-hot oven. When done and cold, spread half of the jelly over one layer, cover with banana slices, lay over the second layer, put on the remaining jelly and bananas; mix the whipped cream with 1 tablespoonful fine sugar and a little vanilla, cover the whole cake with cream, or take 1 pint of whipped cream and put half of the cream between the layers and the remaining over the top, and serve.

Neapolitan Cake. — Roll out some puff paste to ⅛ inch in thickness, cut it into 3 strips 5 inches wide and about 10 inches long; moisten a large shallow tin pan with cold water, put in the strips, dust them with powdered sugar, and bake in a medium-hot oven. When done and cold, cover 1 strip with boiled vanilla cream (see recipe No. 139), put over this the second strip, and spread over some currant jelly; lay on the third strip. Mix ½ cup powdered sugar with 1 tablespoonful boiling water and a few drops of lemon juice, pour it over the cake, and set aside till firm.

285

One-Egg Cake. — One cup sugar, 1 egg, a piece of butter the size of a walnut, 2 scant cups of flour sifted with 1 heaping teaspoonful baking powder, ½ teaspoonful extract of lemon or vanilla, 1 cup of milk; rub butter and flour together, add the sugar, milk, and egg; mix into a batter, butter a square pan, dust with flour, put in the mixture, and bake in a medium-hot oven till done. This mixture may be baked in 3 small jelly tins, and when done lay them over one another with jelly marmalade or cream between them; or bake it in a pan 12 inches long and 8 inches wide and 1½ inch deep. When done, cut the cake in half, lay them over one another with jelly or cream between, then mix 1 cup sifted powdered sugar with 1 teaspoonful lemon juice and 2 tablespoonfuls boiling water, stir until smooth, pour the icing over the cake, and let stand till firm.

For a chocolate cake, bake the cake the same way, then mix 1 cup of powdered sugar with the white of 1 egg; melt 4 ounces Baker's chocolate, add it to the sugar, mix all together, put half of it between the cake, and spread the remaining over the top of the cake.

For a strawberry shortcake, bake the mixture in 2 small well buttered and floured jelly tins, wash and mash 1 quart of strawberries, mix with ½ cup sugar, put half of them between the 2 layers, and the remainder on top; serve with cream or vanilla sauce, or put some whipped cream over the strawberries.

Spice Cake.— Three fourths cup butter, 1 cup molasses, 1 cup sugar, 3 eggs, 3 cups flour sifted with 1½ teaspoonful baking powder, 1½ teaspoonful cinnamon, 1 teaspoonful cloves, ½ grated nutmeg, 1 cup sour milk or cold coffee. Stir butter and sugar to a cream, add the eggs one at a time, stir a few minutes between each addition, add molasses and spice, then alternately flour and milk. Butter a square cake pan, dust with flour, pour in the cake mixture, and bake in medium-hot oven; or bake small cakes in gem pans and when cold ice them with sugar glaze.

Molasses Cake.— One cup molasses, ½ cup butter, 2 eggs, ½ cup milk, ½ tablespoonful ginger, 2 cups flour sifted with 1 teaspoonful baking powder. Mix and bake the same as above.

286

Gingerbread.— One cup brown sugar, 1 cup molasses, ½ cup butter or lard, 2 eggs, 1 tablespoonful ginger, ½ teaspoonful ground mace, ½ cup cold coffee, ½ teaspoonful salt, 1 heaping teaspoonful baking powder sifted with 3 cups of flour. Stir butter or lard with the sugar to a cream, add the eggs one at a time, stirring a few minutes between each addition; then add the spice and syrup, last the flour and coffee alternately; pour the mixture into a square or long pan previously well buttered and dusted with flour; bake in a medium-hot oven.

Ginger Snaps.— Half pound brown sugar, ½ pound butter, 1 pint molasses, 1 tablespoonful ginger, 1 teaspoonful cloves, 2 teaspoonfuls cinnamon, 1 teaspoonful baking powder sifted with 2 cups flour. Stir butter and sugar to a cream, add the molasses and spice; when well mixed add flour and work it into a soft dough; if necessary, add more flour, roll out very thin, cut into rounds, and bake on buttered tins.

Corn Bread.— One pint of corn meal, ½ cup of flour, 1 teaspoonful baking powder, 1 egg, ½ teaspoonful salt, 1½ tablespoonful sugar, 1 cup milk. Mix all together and bake in a well-buttered square tin pan. This bread should be about 1½ inch thick when done.

FROZEN DESSERTS.

Mignon Cream. — One pint milk, the yolks of 4 eggs, ½ cupful cream, 1 cupful sugar, 1 teaspoonful vanilla sugar. Place a small pan with 3 tablespoonfuls of the sugar over the fire, stir until it melts and turns light brown, then add ¼ cup hot water, let it boil to a thin syrup and add it to the milk; add the yolks, sugar, and vanilla, mix well and freeze.

Malborn Cream. — Cut ½ pound candied fruit into small pieces, place it in a bowl with ½ gill of sherry wine, then place a double boiler with ½ pint of milk, the yolks of 2 eggs, and ¾ cup of sugar over the fire, stir until just about to boil, remove instantly, and when cold add ½ pint of cream, 1½ tablespoonful best rum, put the cream in a freezer, and freeze till nearly stiff, then add the candied fruit; continue the freezing till firm.

287

Banana Ice Cream. — Remove the peel from 6 ripe bananas and mash them fine, mix 1 pint of cream with ½ pint milk, 1 cup of sugar, and 2 teaspoonfuls vanilla; put the cream into the freezer, and freeze till it begins to thicken, then add the bananas, and freeze till firm.

Frozen Caramel Cream. — Put in a double boiler 1 cup milk, 1 cup cream, ¾ cup sugar, and the yolks of 3 eggs; beat well, then place over the fire, and stir till nearly boiling; when cold, add this slowly to the 3 beaten whites while stirring constantly; put 3 tablespoonfuls sugar over the fire, stir till it turns yellow, add a little water, boil to a syrup. When cold, add it to the cream, and freeze. In place of caramel, 1 teaspoonful vanilla extract may be added.

Ice Cream without Milk or Cream. — One pint water, 1 ounce butter, the yolks of 3 eggs, 1 cup of sugar, the peel of ½ lemon, and ½ tablespoonful lemon juice; beat 3 whites to a stiff froth; wash the butter several times, stir sugar and yolks to a cream, add the water and butter, place it in a double boiler, stir till nearly boiling; remove, and when cold add the whites, then freeze.

Lemon Sherbet.— The juice of 4 lemons and 1 orange, 2½ cups sugar, 1 quart water; boil sugar and water to a syrup; when cold add the lemon and orange juice and freeze; add last 1 or 2 beaten white of eggs, mix, and serve.

Strawberry Sherbet.— Put the juice of 1 lemon over 1 quart of mashed strawberries; boil 1 quart water with 2 cups sugar, pour it over the strawberries, and when cold rub them through a sieve, then freeze; add last the white of 1 beaten egg, mix, and serve.

Coffee Frappe.— Boil 1 quart water with ½ cup sugar, add 4 ounces fine-ground coffee, cover and set on side of the stove 10 minutes; then strain, and when cold add the white of 1 egg; then freeze, and serve in glasses with whipped cream on top.

Coffee Sorbet.— Pour 3 pints of boiling water over 1 cup of fine-ground coffee, cover, and let it stand 15 minutes, then strain through a napkin; add 1 cup sugar, stir till dissolved, and when cold freeze it till nearly stiff; add 1 gill of the best brandy, continue the freezing for a few minutes, and serve.

CUSTARD.

Caramel Custard.— Boil 1 cup sugar with ½ cup water till the sugar begins to turn light brown, then pour it into a pudding-dish. Mix at the same time 1 quart of milk with 6 eggs, 4 tablespoonfuls sugar, 1 teaspoonful vanilla extract; pour this into the dish, set the dish in a pan of water, and bake till the custard has set. Remove and place it for several hours on ice. In serving, turn the custard out on to a dish, and serve. This custard may be put into small molds or cups and baked the same way.

Cocoanut Caramel.— Mix the whites of 8 eggs with 1½ pint of milk, ½ cup sugar, 1 teaspoonful vanilla extract, and 1½ cup fresh cocoanut; let it stand 1 hour. Place 1 cup of sugar with ½ cup water over the fire, boil until it begins to turn yellow, then pour it into 6 small bowls; spread the caramel even with a spoon so that the bowls are completely lined inside, then pour in the custard, set the bowl in a pan of water so that the water reaches halfway up the bowls, and bake till the custard is firm to the touch. When done, remove the bowls and set them in a cool place. In serving, turn the custard on to small plates, and serve.

Caramel Charlotte. — Put ½ ounce gelatin in a small bowl with ½ gill of cold water; at the same time place a small saucepan with 3 tablespoonfuls sugar over the fire, stir until the sugar is melted and has assumed a rich brown color, then add ½ pint milk; cook and stir till the sugar is dissolved, mix the yolks of 5 eggs with 3 tablespoonfuls sugar, 1 teaspoonful vanilla, and 2 tablespoonfuls cold milk; add a little of the hot caramel milk to the yolks, then add the yolks to the caramel milk, stir over the fire till nearly boiling; instantly remove, add the gelatin, stir until dissolved, then strain into a bowl, and set aside to cool. Beat 1 pint of cream till stiff, when the caramel mixture begins to thicken add it slowly to the cream while beating constantly; in the meantime fit a piece of white paper in the bottom of a charlotte mold, line the sides and bottom with thin slices of sponge cake, pour in the cream, cover 289 the top with sponge cake the same way. Place the charlotte on ice for several hours. When ready to serve turn the charlotte on to a dish and garnish with a wreath of spun sugar or serve plain.

Strawberry Charlotte. — Prepare a strawberry fromage, No. 189, line a mold with sponge cake or lady fingers the same as in the foregoing recipe, pour in the strawberry fromage, cover with the same cake or fingers, and set on ice. When ready to serve, turn the charlotte on to a dish, remove the paper, and serve with cream, which should be sweetened with sugar and flavored with vanilla, or serve plain without the cream. In place of strawberry fromage any other kind of fromage may be used.

SAUCES.

Bismarck Sauce. — Stir the yolk of 2 eggs with 1 cup of powdered sugar to a cream, add slowly ½ cup of Rhine wine, beat the white to a stiff froth, add the sauce slowly to the white while beating constantly, add last ½ cupful whipped cream. In place of Rhine wine sherry wine may be taken.

Transparent Sauce. — Mix 1 heaping teaspoonful cornstarch in a small saucepan with ¼ cup cold water, add 1 cupful boiling water and the thin peel of ½ lemon; set the saucepan over the fire, stir and boil a few minutes, then remove, add 2 tablespoonfuls lemon juice and 3 or 4 tablespoonfuls sugar, or sweeten to taste. To this sauce a few spoonfuls of strawberry or raspberry syrup or juice may be added.

Orange Sauce.— Stir the yolks of 3 eggs with 1 cupful powdered sugar to a cream, add slowly 1 cupful orange juice and 3 tablespoonfuls lemon juice; beat the whites to a stiff froth, add slowly while beating constantly the orange mixture to the whites; serve either with hot or cold puddings.

Cream Sauce.— Stir the yolks of 2 eggs with ½ cupful powdered sugar to a cream, add ½ cupful orange juice and 1 tablespoonful lemon juice; beat the whites to a stiff froth, add the orange mixture slowly to the whites while beating constantly; add last 1 cupful whipped cream. In place of orange juice any kind of fruit 290 juice may be taken, or jelly may be dissolved in hot water and used the same way.

Fruit Sauce.— Mix 1 teaspoonful cornstarch in a small saucepan with ½ gill of cold water, add while stirring constantly 1 cupful boiling water, the thin peel of ½ lemon; place the saucepan over the fire and boil a few minutes; remove from fire, add ¾ cupful fruit syrup, either of raspberry, strawberry, apricot, or cherries, and 1 tablespoonful lemon juice; if handy, add 2 or 3 tablespoonfuls white wine, and serve with soufflé and light delicate puddings. It may also be served cold with puddings. In place of fruit syrup, strawberry or cherry marmalade may be taken, or apple or currant jelly.

Raspberry Sauce.— Mix 1 heaping teaspoonful cornstarch in a small saucepan with a little cold water; add slowly while stirring constantly 1 cupful boiling water, asmall piece of cinnamon, and the thin peel of ½ lemon; place the saucepan over the fire, cook a few minutes, then remove, add 1 cupful fresh raspberry juice, ½ cupful sugar, and 1 tablespoonful lemon juice; if liked and handy, alittle Rhine wine or white wine may be added; serve either hot or cold. If raspberry syrup is used, omit the sugar.

Cream Cakes Glassé.— Boil ½ pint milk with 2 ounces butter, add 4 ounces flour, stir until it forms into a smooth paste and loosens itself from the bottom and sides of the saucepan, transfer the paste to a dish, and when nearly cold add the yolks of 4 eggs, and last the beaten whites; drop this mixture (by tablespoonfuls) on to buttered tins, not too close together, brush them over with the beaten egg, and bake to a fine golden color and well done. When done and cold, cut the cakes open on the side and fill them with vanilla cream, No. 129; half the quantity of cream will be sufficient. Place the cakes on a sieve, boil 1 cup of sugar with ½ cup water till the sugar begins to turn light brown (caramel), instantly remove, and pour it over the cakes.

291

WAR RECIPES

Economical Jelly Roll.— Separate three eggs; to the yolks add half a pound of powdered sugar; beat fifteen minutes; then add grated lemon rind; half a pound of sifted flour with a quarter teaspoonful of baking powder; add it to the mixture with the whites beaten stiff and half a cup of milk.

Spread, dust with sugar, and bake till done.

Millionaire Cake.— Cream the yolks of three eggs with one-half cup powdered sugar for ten minutes, add one-half teaspoonful vanilla and three quarters of a cup of flour sifted with one-fourth of a teaspoonful of baking powder and the beaten whites of the three eggs, butter six small layer cake tins and put in the mixture. Bake in a quick oven ten minutes.

Filling.— Put two tablespoonfuls of chocolate and three tablespoonfuls sugar with one-half cup strong coffee and boil for ten minutes, when almost cold add one-half cup well-washed butter, teaspoon vanilla in small portions; when thick and creamy spread between layers and on top and decorate with candied cherries.

Probasco Cream.— Mix 4 ounces flour with 2 ounces butter, 2 ounces ground almonds, 1 yolk and a little water, 1 tablespoonful sugar to a firm paste; let rest 1 hour, roll out ¼ inch thick, lay a small jelly tin over, cut around it, lay the round piece of paste on a buttered tin, brush over with egg and bake a fine golden color. Add to ½ pint whipped cream, 1 teaspoonful granulated gelatine dissolved in 2 tablespoonfuls milk, flavor with ½ teaspoonful vanilla and sweeten with 2 tablespoonfuls sugar, fill into a round form and set on ice.

Boil 1 pound of chestnuts 20 minutes, remove the outside shells and the brown skin, place the nuts in a saucepan, cover with boiling water, cook till very tender, drain and press through a sieve, add 2 tablespoonfuls 292 cream, 2 tablespoonfuls sugar, a pinch of salt, and ½ tablespoonful melted butter. Set on ice. Shortly before serving turn the cream on the almond cake, put the chestnut puree in a pastry bag with a small tube in the end, press the puree in the form of spaghetti around the cream and cover the cream with a thick layer of pulverized macaroons with sponge sugar, sprinkle over 1 tablespoonful satin sugar, ornament with whipped cream and sprinkle with fine pistachio nuts.

Prune Soufflé. — Wash and soak 1 cup of prunes over night, next day drain, remove pits, and add them to 3 whites of eggs beaten stiff; add 1 teaspoonful vanilla, put in buttered and sugared soufflé dish, set in shallow pan with a little water, set in a medium oven to poach till done, which will take about 40 minutes.

Apricot Whip. — Rub half a cup of apricots after they have been cooked through a sieve, add half a bottle of cream beaten stiff to it and two tablespoons powdered sugar and half a teaspoon vanilla, put in six sherbet glasses and decorate with lady fingers around the edge.

Coffee Parfait. — Put ¼ cup granulated sugar with one tablespoon cornstarch and one pint of milk over the fire and boil five minutes, then add two tablespoons coffee extract and ½ teaspoon vanilla; when cool freeze, put into four parfait glasses and decorate with whipped cream.

Dorothy Royal. — Bake a cup cake mixture in a sheet pan, when done and cool, cut into square pieces and pour over a sauce made as follows: put one cup of sugar with half a cup of milk, atablespoon of butter and one square of chocolate over the fire and cook until thick and creamy, about ten minutes; pour hot over the cake, add a spoonful of whipped cream on every piece and sprinkle chopped walnuts over all.

Peach à la Melba. — Put one tablespoon of vanilla ice cream on a round of sunshine cake, on it lay half a preserved peach and pour over two tablespoons melba syrup made as follows: melt ½ cup raspberry syrup to which has been added one tablespoon fine cut candied cherries and decorate with whipped cream.

Praline Cream. — Put ¼ cup granulated sugar with one tablespoon cornstarch and one pint milk over the fire and boil five minutes, stirring all the time, then add the yolks of two eggs and one teaspoon vanilla 293 and set aside to cool, in another saucepan put ½ cup granulated sugar with ¼ cup water over the fire and boil until it turns a golden brown; pour on a buttered marble and when cold roll with a rolling pin until fine, sprinkle into the cooked mixture, add ½ pint cream and freeze in a well-packed freezer.

Prune Whip.— Rub ½ cup prunes after they have been cooked through a sieve; add ½ bottle whipped cream to it and two tablespoonfuls powdered sugar, ½ teaspoon vanilla, serve in six sherbet glasses with lady fingers around the edge.

War Bread.— Two cups of rye flour, one cup bran, one cup Indian meal, one teaspoonful salt, one yeast cake and 2 cups of luke warm water, two tablespoonfuls molasses, one teaspoonful fat; put flour, bran and meal into a bowl, add the salt and rub the fat through the flour, then add the yeast and crumble it fine, then add the warm water and mix well about 15 minutes, cover and let raise to double its height, then add some wheat flour slowly while kneading the dough to a soft firm dough; shape it into loaves, put into buttered pan, let rise again till double its size, and bake in a medium hot oven till done about ¾ of an hour.

Bran Muffins.— Put one cup of bran and one cup of whole wheat in a bowl, add one teaspoon salt and one teaspoon fat and rub it through until fine and then add one egg, two tablespoons molasses and one cup milk and a heaping teaspoon baking powder, mix well and put in buttered and floured muffin tins and bake about fifteen minutes.

Oatmeal Cookies.— Cream two tablespoons butter with ½ cup sugar until white and creamy, add one egg and stir again a few minutes, then add ½ cup oatmeal and ½ cup flour, two tablespoons raisins, one tablespoon molasses and ¼ cup milk; drop by teaspoons on well-buttered tins and bake in a hot oven about ten minutes.

Vienna Pancakes.— Put two ounces flour with three eggs in a bowl, add ½ teaspoon salt, one cup milk, one teaspoon sugar and beat fifteen minutes, heat a large well-cleaned frying pan, melt ½ tablespoon fat or butter, pour in the mixture, turn the pan from side to side, let the batter run up the sides of the pan, continue until the batter has formed a coating on bottom and sides of the pan, then set in a hot oven and bake until 294 light brown on top; serve dusted with powdered sugar and the juice of ½ lemon.

Checkerboard Sandwiches.— Take an 8-cent white loaf of bread, and an 8-cent loaf of brown bread, cut off crusts on all sides of both loaves; now cut into lengths of 2 inches thick, butter thickly on all sides, lay a white strip next a brown strip of bread alternately to form checkers; then roll in a wet napkin and set aside to chill; when required cut in slices and serve. When finished they should look like checkerboards.

Domino Sandwiches.— Stir one cream cheese with ½ cup cream and two dashes of paprika until smooth; spread on brown bread and cover with brown bread; cut the sandwiches 1½ inches by three inches and decorate top with the cheese mixture put through a pastry bag to represent dominoes.

Minced Ham Sandwiches.— Put cooked ham through meat machine and to one cup of ham add ¼ cup mayonnaise spread between thin pieces of wheat bread cut in diagonal shapes and put a little of the mixture in the center of each sandwich.

Fruit Sandwich.— Cook ½ cup dates and ½ cup figs in water five minutes; drain, chop fine, mix with ½ cup apple or currant jelly and ½ cup of chopped walnuts or pecans. Cut graham bread into round and heart shapes and put mixture between and decorate top with a little of the mixture.

Toasted Cheese Roll.— Sprinkle grated cheese on thin slices of wheat bread that has been buttered, sprinkle a little paprika over it and toast until cheese is melted, holding cheese side to the flame. Then roll quickly, hold together with a toothpick and toast the outside.

Cheese Balls.— Grate one-third of a cup of American cheese, add the beaten whites of one egg to it and a little paprika, drop by teaspoonfuls into bread crumbs and then fry in hot fat a minute till brown; nice served with fish or salad.

Horn of Plenty Salad.— Cut pineapple from top lengthwise cutting across the bottom, remove the fruit of pineapple, cutting it in small pieces, careful not to break the skin; add one pint of fine cut peeled apples, one pint of fine cut celery, ½ cup of walnut meats broken, mix 295 with 1 cup of fine mayonnaise, arrange the pineapple on a platter the green top up and fix it to look like a horn of plenty; fill in the salad, pour over one cup of fine mayonnaise, spread it smooth with a knife, decorate with celery tops, candied cherries and chopped pistachio nuts.

Grape Fruit Salad.— Cut three grape fruit and three oranges in half and take out the pieces, add one pint fine cut celery, four tablespoons fine mayonnaise and put back in the grape fruit shells that have been cleaned out and scalloped. Mask with mayonnaise and garnish with pieces of grape fruit and orange and celery leaves, sprinkle finely chopped pistachio nuts over all.

Waldorf Salad.— Peel, core and cut fine, four apples, add one pint finely cut celery to it and mix two tablespoonfuls broken English walnuts previously boiled in salted water, add four tablespoonfuls fine mayonnaise and mix well, pile on a platter and mask with mayonnaise, garnish with pieces of red apple, celery leaves and walnuts.

Pear farcis fromage à la Rocquefort.— Peel a nice mellow ripe pear, remove core after cutting it carefully in half, cream, put Rocquefort cheese through a potato ricer, add little cream and a few dashes of paprika, mix few minutes, fill in center of pears, lay the pear on a crisp lettuce leaf, decorate with chopped pistachio nuts, and a candied cherry at top, pour over a nice Russian dressing, and serve.

Russian Dressing.— Put 2 yolks of eggs in a bowl, set the bowl in cracked ice or cold water, add to the yolks ¼ teaspoonful dry mustard, add ¼ cup of salad oil, in small portions till absorbed by the yolks, add 1 teaspoonful lemon juice or vinegar, 1 even teaspoonful salt, dash of paprika, 1 tablespoonful Chili sauce, and 1 tablespoonful whipped cream.

Green Pepper farcis fromage à la Cream.— Have a washed green pepper scooped out so the seeds are all removed. Mix a fresh cream cheese with fine chopped walnuts and pieces of pimento cut into small squares, fill into the pepper, so it is well packed, set on ice till chilled an hour or so.

When ready to serve, cut into slices with a sharp knife, arrange onto fresh lettuce leaves, pour over French dressing and serve.

297

CONTENTS.

Recipe numbers and *italicized* items were added by the transcriber. All headers were printed as shown; where there is no corresponding header in the body text, links lead to the first numbered recipe. Recipes in the Appendix are unnumbered.

61.	Coffee Cream Sauce	15
62.	Nutmeg Sauce	15
63.	Orange Cream Sauce	15
64.	Sabayon Sauce	15
65.	Strawberry Chaudeau Sauce	16
66.	Pineapple Chaudeau Sauce	16
67.	Raspberry Chaudeau Sauce	16
68.	Cocoanut Snow Sauce	16
69.	Cocoanut Sauce (another way)	16
70.	Snow Sauce with Orange Flavor	16
71.	Pistachio Sauce	17
72.	Cold Pineapple Sauce	17

FRUIT AND SUGAR SYRUPS.

73.	Plain or Sugar Syrup	17
74.	Pineapple Syrup	17
75.	Strawberry Syrup	17
75a.	Raspberry Syrup	18
76.	Raspberry and Currant Syrup	18
77.	Raspberry Syrup without Fruit	18
78.	Raspberry Syrup without Boiling	18
79.	Blackberry Syrup	18
80.	Peach Syrup	19
81.	Apricot Syrup	19
82.	Cherry Syrup	19
83.	Wild Cherry Syrup	19
84.	Wild Cherry Bark Syrup	19
85.	Vanilla Syrup	19
86.	Vanilla Cream Syrup	19
87.	Cream Syrup	20
88.	Lemon Syrup	20
29989.	Lemon Syrup with Oil of Lemon	20

FINE COLD PUDDINGS.

BOILED AND BAKED PUDDINGS.

RICE PUDDINGS AND DISHES MADE OF RICE
FOR DESSERT.

COLD PUDDINGS MADE WITH MILK.

STRUDEL, STRAWBERRY SHORTCAKES, BABA, SOLEIL, BRIOCHE, SAVARIN AND DAMPFNUDELN.

ROLLS AND BREAD.

APPENDIX.

FROZEN DESSERTS.

ALPHABETICAL INDEX.

This Index was added by the transcriber. Recipes in the main text are identified by number, those in the Appendix by page.

A B C D E F G H I J K L M
N O P Q R S T U V W Z

316